# Large-Scale Graph Processing
# Using Apache Giraph

Sherif Sakr · Faisal Moeen Orakzai
Ibrahim Abdelaziz · Zuhair Khayyat

# Large-Scale Graph Processing Using Apache Giraph

Springer

Sherif Sakr
School of Computer Science
and Engineering
University of New South Wales
Sydney, NSW
Australia

Faisal Moeen Orakzai
Department of Computer Science
Aalborg University
Aalborg
Denmark

Ibrahim Abdelaziz
King Abdullah University of Science
and Technology
Thuwal
Saudi Arabia

Zuhair Khayyat
King Abdullah University of Science
and Technology
Thuwal
Saudi Arabia

ISBN 978-3-319-83735-2          ISBN 978-3-319-47431-1    (eBook)
DOI 10.1007/978-3-319-47431-1

Printed on acid-free paper

This Springer imprint is published by Springer Nature
The registered company is Springer International Publishing AG
The registered company address is: Gewerbestrasse 11, 6330 Cham, Switzerland

*To my wife, Radwa, my daughter, Jana, and my son, Shehab for their love, encouragement, and support.*

Sherif Sakr

*To my father, Moeen Orakzai, my mother, Hikmat Moeen, my wife, Sana and my son Ibrahim for their support.*

Faisal Moeen Orakzai

*To my wife and my lovely daughters for their unconditional love and support.*

Ibrahim Abdelaziz

*To my father, Yarub, my mother, Nadia, my lovely wife, Manal, and my two handsome boys, Mazin and Ziyad, for their love and support.*

Zuhair Khayyat

# Foreword

The present decade has been dubbed the Digital Universe Decade because the digital universe, all data available in digital form, is growing at an exponential rate during this decade. Lots of this data is graph data. Examples include data related to social network connections and data diffusion, computer networks, telecommunication networks, the Web, and knowledge bases. Yet another example close to my heart is road-network data: in step with the rapid increases is available vehicle trajectory data, this data grows rapidly in size, resolution, and sophistication.

Graph analytics enables the extraction of valuable information from graph data and enables valuable services on top of graph data. For example, in social media, graph analytics can be used to detect communication patterns that might be of interest to national defense. In the medical and biological domains, graph analytics may be used for analyzing relationships in the contexts of proteins, DNA, cells, chemical pathways, and organs in order to determine how they are affected by lifestyle choices and medications. In transportation, graph analytics can enable personalized and stochastic routing that takes into account time-varying travel times and greenhouse gas emissions. These are but some uses of graph data analytics; new ones emerge at an accelerated pace.

Practical graph analytics uses a combination of graph-theoretic, statistical, and data management techniques to model, store, query, and analyze graph data. The processing of graph data embodies computationally hard tasks that have only been exacerbated by the rapid and increasing growth in available graph-related data. In 2010, Google introduced Pregel, a distributed system for large-scale graph processing. Inspired by the Bulk Synchronous Parallel model, Pregel features an intuitive, vertex-centric organization of computations that lets its users "think like a vertex". Since its introduction, Pregel has spurred substantial interest in large-scale graph data processing, and a number of Pregel-like systems have emerged.

Apache Giraph is an open-source system for Pregel-like, large-scale graph data processing. It has a global and growing user community and is thus an increasingly popular system for managing and analyzing graph data. This book provides a step-by-step guidance to data management professionals, students, and researchers who are looking to understand and use Apache Giraph. It guides the reader through the details of installing and configuring the system. It offers a detailed description of the programming model and related aspects. And it offers a step-by-step coverage of the implementation of several popular, as well as advanced, graph analytics algorithms, covering also related optimization details.

In a nutshell, this book is a timely and valuable resource for data management professionals, students, and researchers interested in using Apache Giraph for large-scale graph processing and graph analytics.

Aalborg, Denmark                                                                    Christian S. Jensen
July 2016

# Preface

We are generating data more than ever. The ubiquity of the Internet has dramatically changed the size, speed, and nature of the generated data. Almost every human became a data generator and every business became a digital business. As a result, we are witnessing a data explosion. In the past few years, several technologies have contributed to this data explosion including mobile computing, Web 2.0, social media, social network, cloud computing and Software-as-a-Service (SaaS). In the future, it is expected that the Internet of Things will further amplify this challenge. In particular, several *things* would be able to get connected to the Internet, and thus there will be lots of data passed from users to devices, to servers, and back. Hence, in addition to the billions of people who are currently using the Internet and daily producing a lot of data, watches, cars, fridges, toaster, and many other devices will be online and continuously generating data as well. It is quite expected that in the near future, our toasters will be able to recommend types of bread based on suggested information from our friends on the social networks.

With the recent emerging wave of technologies and applications, the world has become more connected than ever. Graph is a popular neat data structure which is used to model the data as an arbitrary set of objects (vertices) connected by various kinds of relationships (edges). With the tremendous increase in the size of the graph-structured data, large-scale graph-processing systems have been crucially on demand and attracted a lot of interest. This book is intended to take you to a journey with **Apache Giraph**, a popular distributed graph-processing platform, which is designed to bring the power of big data processing to graph data that would be too large to fit on a single machine. We describe the fundamental abstractions of the system and its programming models and describe various techniques for using the system to process graph data at scale. The book is designed as a self-study step-by-step guide for any reader with an interest in large-scale graph processing. All the source codes presented in the book are available for download from the associated Github repository of the book.

## Organization of the Book

Chapter 1 starts with a general background of the big data phenomena. We then introduce the big graph problem, its applications, and how it differs from the traditional challenges of the big data problem and motivates the need for domain-specific systems that are designed to tackle the large-scale graph-processing problem. We then introduce the Apache Giraph system, its abstraction, programming model, and design architecture to set the stage for the reader and provide him with the fundamental information which is required to smoothly follow the other chapters of the book.

Chapter 2 takes Giraph as a platform. Keeping in view that Giraph uses Hadoop as its underlying execution engine, we explain how to set up Hadoop in different modes, how to monitor it, and how to run Giraph on top of it using its binaries or source code. We then move to explaining how to use Giraph. We start by running an example job in different Hadoop modes and then approach more advanced topics such as monitoring Giraph application life cycle and monitoring Giraph jobs using different methods. Giraph is a very flexible platform and its behavior can be tuned in many ways. We explain the different methods of configuring Giraph and end the chapter by giving a detailed description of setting up a Giraph project in Eclipse and IntelliJ IDE.

Chapter 3 provides an introduction to Giraph programming. We introduce the basic Giraph graph model and explain how to write a Giraph program using the vertex similarity algorithm as a use case. We explain three different ways of writing the driver program and their pros and cons. For loading data into Giraph, it comes packaged with numerous input formats for reading different formats of data. We describe each of the formats with examples and end the chapter with the description of Giraph output formats.

Chapter 4 discusses the implementation of some popular graph algorithms including PageRank, connected components, shortest paths, and triangle closing. In each of these algorithms, we give an introductory description and show some of its possible applications. Then using a sample data graph, we show how the algorithm works. Finally, we describe the implementation details of the algorithm in Giraph.

Chapter 5 sheds light on the advanced Giraph programming. We start by discussing common Giraph algorithmic optimizations and how those optimizations may improve the performance and flexibility of the algorithms implemented in Chap. 4. We explain different graph optimizations to enable users to implement complex graph algorithms. Then, we discuss a set of tunable Giraph configurations that controls Giraph's utilization of the underlying resources. We also discuss how to change Giraph's default partitioning algorithm and how to write a custom graph input and output format. We then talk about common Giraph runtime errors and finalize the chapter with information on Giraph's failure recovery.

Recently, several systems have been introduced to tackle the challenge of large-scale graph processing. In Chap. 6, we highlight two of these systems, GraphX and GraphLab. We describe their program abstractions and their programming models. We also highlight the main commonalities and differences between these systems and Apache Giraph.

## Target Audience

We hope this book serves as a useful reference for students, researchers, and practitioners in the domain of large-scale graph processing.

**To Students**: We hope that the book provides you an enjoyable introduction to the field of large-scale graph processing. We have attempted to properly describe the state of the art and present the technical challenges in depth. The book will provide you with a comprehensive introduction and hands-on experience to tackling large-scale graph-processing problem using the Apache Giraph systems.

**To Researchers**: The material of this book will provide you with a thorough coverage for the emerging and ongoing advancements on big graph-processing systems. You also can use this book as a starting point to tackle your next research challenge in the domain of large-scale graph processing.

**To Practitioners**: You will find this book a very useful step-by-step guide with several code examples, with source codes available in the Github repository of the book, and programming optimization techniques so that you can immediately put the gained knowledge from this book into practice due to the open-source availability of Apache Giraph system.

Sydney, Australia                                                          Sherif Sakr
Aalborg, Denmark                                                  Faisal Moeen Orakzai
Thuwal, Saudi Arabia                                              Ibrahim Abdelaziz
Thuwal, Saudi Arabia                                                Zuhair Khayyat

# Acknowledgments

I would like to thank my parents for their encouragement and support. I want to thank my children, Jana and Shehab, for the happiness and enjoyable moments they are always bringing to my life. My most special appreciation goes to my wife, Radwa Elshawi, for her everlasting support and deep love.

Sherif Sakr

I want to thank my father Moeen, for allowing me to follow my ambitions throughout my life and my mother Hikmat for wishing the best for me at every step. I want to thank Sana, for her constant love and support in spite of all the time it kept me away and to my son, Ibrahim, a recent addition to the family who is always making me smile.

Faisal Moeen Orakzai

I am filled with gratitude to my wife for her encouragement, dedication, and support. I am also grateful to my daughters for their love and the great time I spend with them.

Ibrahim Abdelaziz

I would like to thank my parents, Yarub and Nadia, for their endless encouragement and support. I would like to also thank my once-in-a-life-time love, Manal, for standing beside me throughout my studies and writing this book. Special thanks to my two kids, Mazin and Ziyad, for their enthusiasm and keeping me awake at night.

Zuhair Khyyat

# Contents

# About the Authors

**Sherif Sakr** is currently Professor of Computer and Information Science in the Health Informatics Department at King Saud bin Abdulaziz University for Health Sciences. He is also affiliated with the University of New South Wales and DATA61/CSIRO (formerly NICTA). He received his Ph.D. degree in Computer and Information Science from Konstanz University, Germany in 2007. He received his B.Sc. and M.Sc. degrees in Computer Science from the Information Systems Department at the Faculty of Computers and Information in Cairo University, Egypt, in 2000 and 2003, respectively. In 2008 and 2009, Sherif held an Adjunct Lecturer position at the Department of Computing of Macquarie University. He also held visiting appointments in several academic and research institutes including Microsoft Research (2011), Alcatel-Lucent Bell Labs (2012), Humboldt University of Berlin (2015), University of Zurich (2016), and TU Dresden (2016). In 2013, Sherif has been awarded the Stanford Innovation and Entrepreneurship Certificate.

**Faisal Moeen Orakzai** is a joint Ph.D. candidate at Université Libre de Bruxelles (ULB) Belgium and Aalborg University (AAU) Denmark. He received his joint Masters degree in Computer Science from ULB and TU Berlin in 2014. After graduating as a computer engineer in 2006 from National University of Sciences and Technology (NUST), he worked in the enterprise software development industry for a couple of years before getting attracting towards distributed and large-scale data processing systems. In addition to research, he works as a consultant and helps companies setting up their distributed data processing architectures and pipelines. He is a big data management and analytics enthusiast and currently working on a Giraph based framework for spatio-temporal pattern mining.

**Ibrahim Abdelaziz** is a Computer Science Ph.D. candidate at King Abdullah University of Science and Technology (KAUST). He received his B.Sc. and M.Sc. degrees from Cairo University in 2007 and 2012, respectively. Prior to joining KAUST, he used to work on pattern recognition and information retrieval in several research organizations in Egypt. His current research interests include data mining over large-scale graphs, distributed systems, and machine learning.

**Zuhair Khayyat** is a Ph.D. candidate focusing on big data, analytics, and graphs in the InfoCloud group at King Abdullah University of Science and Technology (KAUST). He also received his Masters degree in Computer Science from KAUST in 2010. He received his B.Sc. degree in Computer Engineering in 2008 from King Fahd University of Petroleum and Minerals (KFUPM).

# List of Figures

# List of Tables

# Introduction

## 1.1 Big Data Problem

There is no doubt that we are living the era of big data where we are witnessing radical expansion and integration of digital devices, networking, data storage and computation systems. We are generating data more than ever. In practice, data generation and consumption is becoming a main part of people's daily life especially with the pervasive availability and usage of Internet technology and applications. The number of Internet users reached 2.27 billion in 2012. As result, we are witnessing constantly the creation of digital data from various sources and at ever-increasing data. Social networks, mobile applications, cloud computing, sensor networks, video surveillance, GPS, RFID, Internet of Things (IoT), imaging technologies, and gene sequencing are just examples of technologies that facilitate and accelerate the continuous creation of massive datasets that must be stored and processed. For example, in 1 min on the Internet, Facebook records more than 3.2 million likes, stores more than 3.4 million posts and generates around 4 GB of data. In March 2013, Facebook launched a graph search feature that enables its users to search the social graph for users with similar locations or hobbies. Also in 1 min, Google answers about 300K searches, 126 h uploaded to YouTube and more than 140K video views, about 700 users created in Twitter and more than 350K tweets generated, and more than 11K searches on LinkedIn performed. These numbers, which are continuously increasing, provide a perception of the massive data generation, consumption, and traffic which are happening in the Internet world. While more and more people will gain access to such a global information and communication infrastructure, big leaps on the Internet traffic, and the amount of generated data are expected.

In another context, powerful telescopes in astronomy, particle accelerators in physics, and genome sequencers in biology are putting massive volumes of data into the hands of scientists. The cost of sequencing one human genome has fallen from $100 million in 2001 to $10K in 2011. Every day, the survey Telescope [2] generates on the order of 30 TB of data, the New YorkStock Exchange captures around 1

© Springer International Publishing AG 2016
S. Sakr et al., *Large-Scale Graph Processing Using Apache Giraph*,
DOI 10.1007/978-3-319-47431-1_1

TB of trade information, and about 30 billion radio-frequency identification (RFID) tags are created. Add to this mix, the data generated by the hundreds of millions of GPS devices are sold every year, and more than 30 million networked sensors are currently in use (and growing at a rate faster than 30 % per year). These data volumes are expected to double every 2 years over the next decade. According to IBM, we are currently creating 2.5 quintillion bytes of data every day.[1] IDC predicts that the worldwide volume of data will reach 40 zettabytes by 2020[2] where 85 % of all of this data will be of new data types and formats including server logs and other machine generated data, data from sensors, social media data, and many more other data sources. Clearly, many application domains are facing major challenges on processing such massive amount of generated data from different sources and in various formats. Therefore, almost all scientific funding and government agencies introduced major strategies and plans to support big data research and applications.

In the enterprise world, many companies continuously collect large datasets that record customer interactions, product sales, results from advertising campaigns on the Web, and other types of information. In practice, a company can generate up to petabytes of information in the course of a year: webpages, clickstreams, blogs, social media forums, search indices, email, documents, instant messages, text messages, consumer demographics, sensor data from active and passive systems, and more. By many estimates, as much as 80 % of this data is semi-structured or unstructured. In practice, it is typical that companies are always seeking to become more nimble in their operations and more innovative with their data analysis and decision-making processes. And they are realizing that time lost in these processes can lead to missed business opportunities. The core of the data management challenge is for companies to gain the ability to analyze and understand Internet-scale information just as easily as they can now analyze and understand smaller volumes of structured information.

The term *Big Data* has been coined under the tremendous and explosive growth of the world digital data which is generated from various sources and in different formats. In principle, the big data term is commonly described by 3V main attributes (Fig. 1.1): the *Volume* attributes describes the massive amount of data that can be billions of rows and millions of columns, the *Variety* attribute represents the variety on formats, data sources, and structures, and the *Velocity* attribute reflects the very high speed on data generation, ingestion, and near real-time analysis. In January 2007, Jim Gray, a pioneer database scholar, described the big data phenomena as the *Fourth Paradigm* [3] and called for a *paradigm shift* in the computing architecture and large-scale data processing mechanisms. The first three paradigms were *experimental, theoretical* and, more recently, *computational science*. Gray argued that the only way to cope with this paradigm is to develop a new generation of computing tools to manage, visualize, and analyze the data flood. According to Gray, computer architectures have become increasingly imbalanced where the latency gap between multi-core CPUs and mechanical hard disks is growing every year which makes the

---

[1]http://www-01.ibm.com/software/data/bigdata/what-is-big-data.html.
[2]http://www.emc.com/about/news/press/2012/20121211-01.htm.

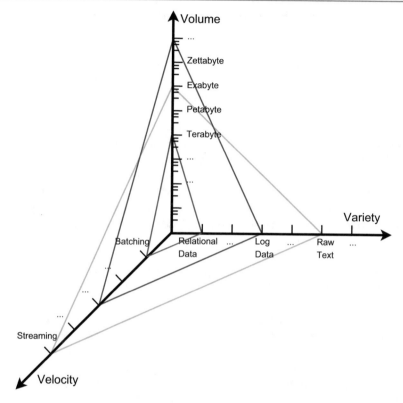

**Fig. 1.1**  3V characteristics of big data

challenges of data-intensive computing much harder to overcome [4]. Hence, there is a crucial need for a systematic and generic approach to tackle these problems with an architecture that can also scale into the foreseeable future. In response, Gray argued that the new trend should instead focus on supporting cheaper clusters of computers to manage and process all this data instead of focusing on having the biggest and fastest single computer. In addition, the 2011 McKinsey global report described big data as the next frontier for innovation and competition [5]. The report defined big data as "*Data whose scale, distribution, diversity, and/or timeliness require the use of new technical architectures and analytics to enable insights that unlock the new sources of business value.*" This definition highlighted the crucial need for new data architecture solution that can manage the increasing challenges of big data problems. In response, the new scale of *big data* has been attracting a lot of interest from both the research and industrial worlds aiming to create the best means to process and analyze this data and make the best use of it.

## 1.2   A Star Is Born: The MapReduce/Hadoop Framework

In our modern world, data is a key resource. However, in practice, data are not useful in and of themselves. They only have utility if meaning and value can be extracted from them. Therefore, given their utility and value, there are always continuous increasing efforts devoted to producing and analyzing them. In principle, big data discovery enables data scientists and other analysts to uncover patterns and corre- lations through analysis of large volumes of data of diverse types. In particular, the power of big data is revolutionizing the world. From the modern business enterprise to the lifestyle choices of today's digital citizen, the insights of big data analytics are driving changes and improvements in every arena. For instance, insights gleaned from big data discovery can provide businesses with significant competitive advan- tages, such as more successful marketing campaigns, decreased customer churn, and reduced loss from fraud. Therefore, it is crucial that all the emerging varieties of data types with huge sizes need to be harnessed to provide a more complete picture of what is happening in various application domains. In particular, in the current era, data represent the new gold while analytics systems represent the machinery that analyzes, mines, models, and mints it.

In 2004, Google made a seminal contribution to the big data world by introduc- ing the MapReduce framework as a simple and powerful programming model that enables easy development of scalable parallel applications to process vast amounts of data on large clusters of commodity machines by scanning and processing large files in parallel across multiple machines [1]. In particular, the framework is mainly designed to achieve high performance on large clusters of commodity machines. The fundamental principle of the MapReduce framework is to move analysis to the data, rather than moving the data to a system that can analyze it. One of the main advantages of this approach is that it isolates the application from the details of run- ning a distributed program, such as issues on data distribution, scheduling, and fault tolerance. Thus, it allows programmers to think in a *data-centric* fashion where they can focus on applying transformations to sets of data records while the details of dis- tributed execution and fault tolerance are transparently managed by the MapReduce framework.

In the MapReduce programming model, the computation takes a set of key/value pairs as input and produces a set of key/value pairs as output. The user of the MapRe- duce framework expresses the computation using two functions: *Map* and *Reduce*. The Map function takes an input pair and produces a set of intermediate key/value pairs. The MapReduce framework groups together all intermediate values associ- ated with the same intermediate key $I$ and passes them to the Reduce function. The Reduce function receives an intermediate key $I$ with its set of values and merges them together. Typically just zero or one output value is produced per Reduce invocation. The main advantage of this model is that it allows large computations to be easily parallelized and re-executed to be used as the primary mechanism for fault tolerance. Figure 1.2 illustrates an example MapReduce program expressed in pseudo-code for counting the number of occurrences of each word in a collection of documents. In this example, the map function emits each word plus an associated mark of occurrences

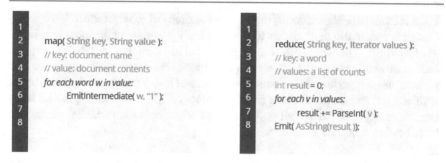

```
1
2    map( String key, String value ):
3    // key: document name
4    // value: document contents
5    for each word w in value:
6        EmitIntermediate( w, "1" );
7
8
```

```
1
2    reduce( String key, Iterator values ):
3    // key: a word
4    // values: a list of counts
5    int result = 0;
6    for each v in values:
7        result += ParseInt( v );
8    Emit( AsString(result ));
```

**Fig. 1.2**  An example MapReduce program [1]

while the reduce function sums together all marks emitted for a particular word. In principle, the design of the MapReduce framework has considered the following main principles [6]:

- *Low-Cost Unreliable Commodity Hardware*: Instead of using expensive, high-performance, reliable symmetric multiprocessing (SMP) or massively parallel processing (MPP) machines equipped with high-end network and storage subsystems, the MapReduce framework is designed to run on large clusters of commodity hardware. This hardware is managed and powered by open-source operating systems and utilities so that the cost is low.
- *Extremely Scalable RAIN Cluster*: Instead of using centralized RAID-based SAN or NAS storage systems, every MapReduce node has its own local off-the-shelf hard drives. These nodes are loosely coupled in rackable systems that are connected with generic LAN switches. These nodes can be taken out of service with almost no impact to still-running MapReduce jobs. These clusters are called Redundant Array of Independent (and Inexpensive) Nodes (RAIN).
- *Fault-Tolerant yet Easy to Administer*: MapReduce jobs can run on clusters with thousands of nodes or even more. These nodes are not very reliable as at any point in time, a certain percentage of these commodity nodes or hard drives will be out of order. Hence, the MapReduce framework applies straightforward mechanisms to replicate data and launch backup tasks so as to keep still-running processes going. To handle crashed nodes, system administrators simply take crashed hardware off-line. New nodes can be plugged in at any time without much administrative hassle. There is no complicated backup, restore, and recovery configurations like the ones that can be seen in many DBMS.
- *Highly Parallel yet Abstracted*: The most important contribution of the MapReduce framework is its ability to automatically support the parallelization of task executions. Hence, it allows developers to focus mainly on the problem at hand rather than worrying about the low- level implementation details such as memory management, file allocation, parallel, multi-threaded, or network programming. Moreover, MapReduce's shared-nothing architecture [7] makes it much more scalable and ready for parallelization.

On the implementation level, the Map invocations of a MapReduce job are distributed across multiple machines by automatically partitioning the input data into a set of $M$ splits. The input splits can be processed in parallel by different machines. Reduce invocations are distributed by partitioning the intermediate key space into $R$ pieces using a partitioning function (e.g., `hash(key) mod R`). The number of partitions ($R$) and the partitioning function are specified by the user. Figure 1.3 illustrates an example of the overall flow of a MapReduce operation which goes through the following sequence of actions:

1. The input files of the MapReduce program is split into $M$ pieces and starts up many copies of the program on a cluster of machines.
2. One of the copies of the program is elected to be the *master* copy while the rest are considered as *workers* that are assigned their work by the master copy. In particular, there are $M$ map tasks and $R$ reduce tasks to assign. The master picks idle workers and assigns each one a map task or a reduce task.
3. A worker who is assigned a map task reads the contents of the corresponding input split and parses key/value pairs out of the input data and passes each pair to the user-defined Map function. The intermediate key/value pairs produced by the Map function are buffered in memory.
4. Periodically, the buffered pairs are written to local disk and partitioned into $R$ regions by the partitioning function. The locations of these buffered pairs on the local disk are passed back to the master, who is responsible for forwarding these locations to the reduce workers.
5. When a reduce worker is notified by the master about these locations, it reads the buffered data from the local disks of the map workers which is then sorted by the intermediate keys so that all occurrences of the same key are grouped together. The sorting operation is needed because typically many different keys map to the same reduce task.
6. The reduce worker passes the key and the corresponding set of intermediate values to the user's Reduce function. The output of the Reduce function is appended to a final output file for this reduce partition.
7. When all map tasks and reduce tasks have been completed, the master program wakes up the user program. At this point, the MapReduce invocation in the user program returns the program control back to the user code.

Figure 1.4 illustrates a sample execution for the example program (`WordCount`) depicted in Fig. 1.2 using the steps of the MapReduce framework which are illustrated in Fig. 1.3. During the execution process, the master pings every worker periodically. If no response is received from a worker within a certain amount of time, the master marks the worker as *failed*. Any map tasks marked *completed* or *in progress* by the worker are reset back to their initial idle state and therefore become eligible for scheduling by other workers. Completed map tasks are re-executed on a failure because their output is stored on the local disk(s) of the failed machine and is therefore inaccessible. Completed reduce tasks do not need to be re-executed since their output is stored in a global file system.

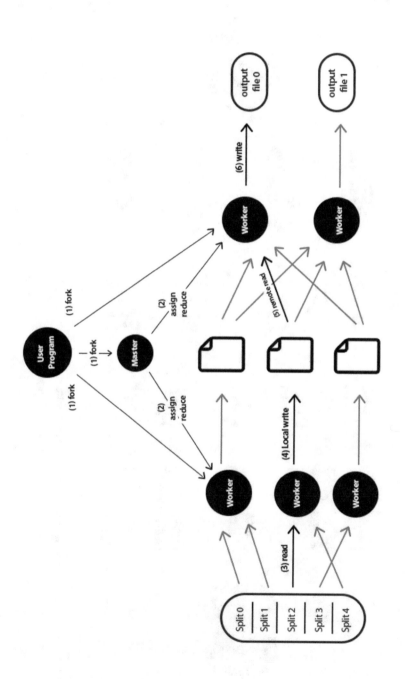

**Fig. 1.3** An overview of the flow of execution a MapReduce operation [1]

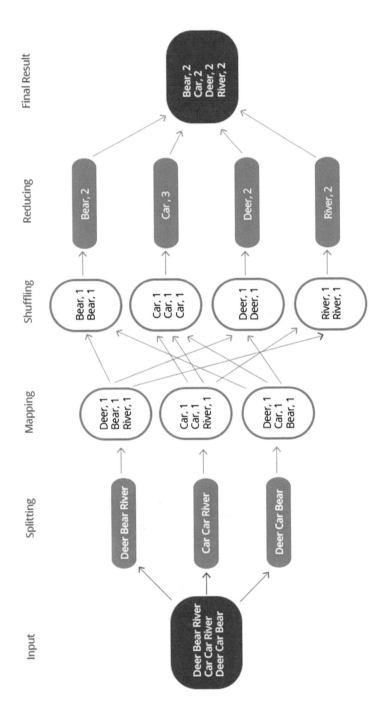

**Fig. 1.4** Execution steps of the word count example using the MapReduce framework

The Apache Hadoop[3] framework has been introduced by Yahoo! as an open-source project [8] that supports data-intensive distributed applications by cloning the design and the implementation of the MapReduce framework.[4] The Hadoop project has been highly successful and created an increasing momentum in the research and business domains. In practice, for about a decade, the Hadoop framework has been recognized as the defacto standard of big data analytics and processing systems. The Hadoop framework was popularly employed as an effective solution that can harness the resources and power of large computing clusters in various application domains [9]. For instance, the Hadoop framework has been widely used by a large number of researchers and business companies in production uses.[5] Due to its wide success, popular technology companies have decided to support the Hadoop framework in their commercial big data processing platforms including *Microsoft*,[6] *IBM*,[7] and *Oracle*.[8] In addition, various emerging and successful startups such as *MapR*,[9] *Cloudera*,[10] *Platfora*,[11] and *Trifcata*[12] have built their solutions and services based on the Hadoop framework. In 2012, Yahoo! declared running the world biggest Hadoop cluster with about 4000 nodes[13] which is used for running various main data processing and analysis job for the company including user logs, advertisement, and financial data. It is predicted that, in the next few years, the global Hadoop market will grow to reach billions of dollars. In addition, around the world, skills related to the Hadoop ecosystem are considered among the top most sought-after skills in the ICT job market.

## 1.3 From Big Data to Big Graphs

In our world, data is not just getting bigger, it is also getting more connected. Recently, people, devices, processes, and other entities have been more connected than at any other point in history. In general, graph is a natural, neat, and flexible structure to model the complex relationships, interactions and interdependencies between objects (Fig. 1.5). In particular, each graph consists of nodes (or vertices) that represent objects and edges (or links) that represent the relationships among the graph nodes.

---

[3] http://hadoop.apache.org/.
[4] In the rest of this book, we use the two names: MapReduce and Hadoop, interchangeably.
[5] http://wiki.apache.org/hadoop/PoweredBy.
[6] http://azure.microsoft.com/en-us/services/hdinsight/.
[7] http://www-01.ibm.com/software/data/infosphere/hadoop/enterprise.html.
[8] http://www.oracle.com/us/products/middleware/data-integration/hadoop/overview/index.html.
[9] https://www.mapr.com/.
[10] http://www.cloudera.com/.
[11] https://www.platfora.com/.
[12] http://www.trifacta.com/.
[13] http://www.informationweek.com/database/yahoo-and-hadoop-in-it-for-the-long-term/d/d-id/1104866?.

**Fig. 1.5** Graph-based
modeling

Graphs have been widely used to represent datasets in a wide range of application domains such as social science, astronomy, computational biology, telecommunications, computer networks, semantic web, protein networks, and many more [10].

The web graph is a dramatic example of a large-scale graph. Google estimates that the total number of web pages exceeded 1 trillion; experimental graphs of the World Wide Web contain more than 20 billion nodes (pages) and 160 billion edges (hyperlinks). Graphs of social networks are another example. Facebook reportedly consists of more than a billion users (nodes) and more than 140 billion friendship relationships (edges) in 2012. In addition, social network graphs are growing rapidly. For instance, Facebook went from roughly 1 million users in 2004 to 1 billion in 2012. In the Semantic Web context, the ontology of DBpedia (derived from Wikipedia), contains 3.7 million objects (nodes) and 400 millions facts (edges).

The ever-increasing size of graph-structured data for these applications creates a critical need for scalable systems that can process large amounts of it efficiently. In practice, graph analytics is an important and effective big data discovery tool. For example, it enables identifying influential persons in a social network, inspecting fraud operations in a complex interaction network and recognizing product affinities by analyzing community buying patterns. However, with the enormous growth in graph sizes, huge amounts of computational power would be required to analyze such massive graph datasets. In practice, distributed and scalable processing of massive graph datasets is a challenging task and have its own specific challenges on top of the general challenges of the big data processing problems. For instance, the iterative nature of graph-processing and computation algorithms typically involve extensive communication and message passing between the graph nodes in each processing step. In addition, graph algorithms tend to be explorative with random access patterns, which are commonly challenging to predict. Furthermore, due to their inherent irregular structure, dealing with the graph partitioning represents a fundamental challenge due to its direct impact on load balancing among the processing nodes, communication cost, and consequently the whole system performance [11].

## 1.4   Use Cases for Big Graphs

Graphs are widely used for data modeling in various application domains such as social networks, knowledge bases, and bioinformatic applications. In this section, we provide an overview of various use cases for graph models.

### 1.4.1   Social Networks

The Web 2.0 [12] and the tremendous growth of Internet applications that facilitate interactive collaboration and information sharing have resulted in the growth of many social networks of different kinds. Nowadays, users tend to share their activities and updates about their daily live with friends via *Facebook*, photos via *Instagram* or *Flicker*, contents via *Twitter*, videos via *YouTube* or *Snapchat*, and professional information via *Linkedin*. The amount of data that is published every day on such social media platforms is becoming increasingly larger. Figure 1.6 illustrates a snippet of an example *directed* social graph while Fig. 1.7 illustrates a snippet of an example *undirected* one. In social networks, nodes are used to represent people or groups while edges used to model the various relationships (e.g., friendship, collaboration, following, business relationship) among them. Due to the increasing sizes of social graphs, Social Network Analysis (SNA) [13] has become one of the crucial tools to investigate the structure of organizations and social groups, focusing on uncovering the structure of the interactions between people. SNA techniques have been effectively used in several areas of interest like social interaction and network evolution analysis, viral marketing and covert networks. For instance, SNA techniques can be used by companies to acquire marketing information from the content or activities from the users of the social networks in order to precisely targeting the advertisement to each user. Collaborative filtering techniques [14] use the user profile information to provide strongly relevant recommendation results.

### 1.4.2   Web Graph

The World Wide Web represents a very large graph [15], where nodes are individual sites or pages and edges are the hyperlinks between pages (Fig. 1.8). In particular, in this graph each vertex is an URL (Unique Resource Locator) and the outgoing edges of a vertex are the hypertext links contained in the corresponding page. The current graph size is estimated at 4.6 billion pages.[14] This graph is a key element for many applications. For example, the basis of Google's search engine is the page

---

[14]http://www.worldwidewebsize.com/.

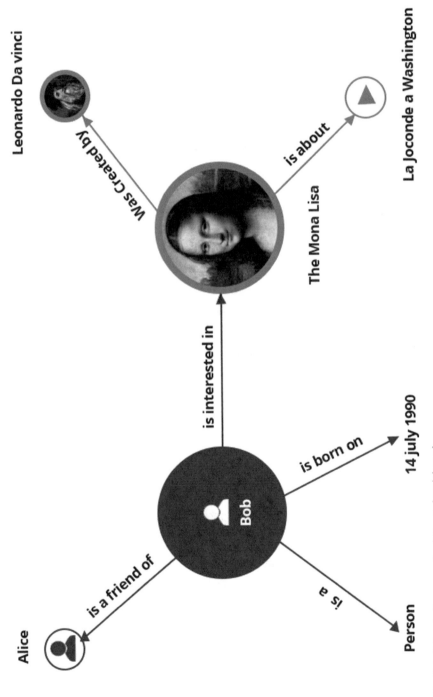

**Fig. 1.6** Snippet of an example of *directed* social graph

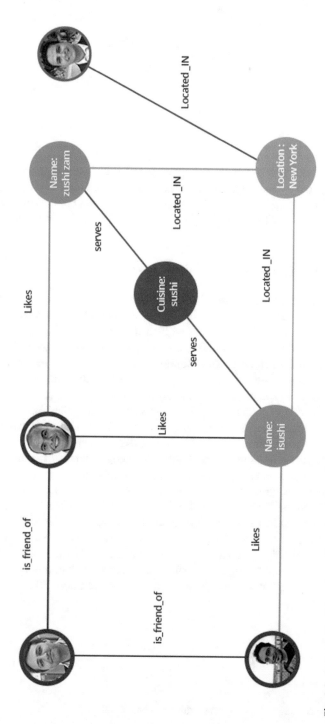

**Fig. 1.7** Snippet of an example *undirected* social graph

**Fig. 1.8**  Structure of Web
Graph

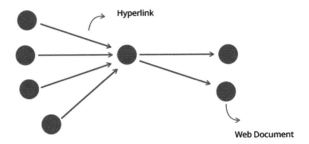

rank algorithm, which determines the importance of a site by the number of sites that
link to it weighted by the importance of those sites [16]. Other applications in the
Web graph include efficiently indexing these Web pages and building search engine
on top of the index will help people quickly find the resources they are looking for.

### 1.4.3  Knowledge Bases and Linked Data

Web knowledge bases are increasingly playing important role in enhancing the intelligence of Web and enterprise search in addition to supporting information integration [17]. In recent years, the Web has evolved from a global information space of
linked documents to a new platform for publishing and connecting structured data on
the Web known as Linked Data. The adoption of Linked Data has lead to the extension of the Web with a global data space that connects data from diverse domains
such as people, books, scientific publications, companies, television, music, and films
(Fig. 1.9). Technically, Linked Data refers to data published on the Web in such a way
that it is machine-readable, its meaning is explicitly defined, it is linked to other external datasets, and can in turn be linked to from external datasets [18]. For example,
Wikipedia has grown into one of the central knowledge sources of mankind, main-
tained by thousands of contributors. The *DBpedia* knowledgebase[15] is a community
effort to extract structured information from Wikipedia and to make this information accessible on the Web [19]. The English version of the DBpedia 2014 dataset
currently describes 4.58 million entities with 583 million facts. In addition, DBpedia provide localized versions in 125 languages where all these versions together
describe 38.3 million things, out of which 23.8 million overlap (are interlinked) with
concepts from the English DBpedia.[16] This huge knowledge-based graph represents
a significant resource for rich inference and reasoning operation [20].

---

[15]http://wiki.dbpedia.org/.
[16]http://wiki.dbpedia.org/services-resources/dbpedia-data-set-2014.

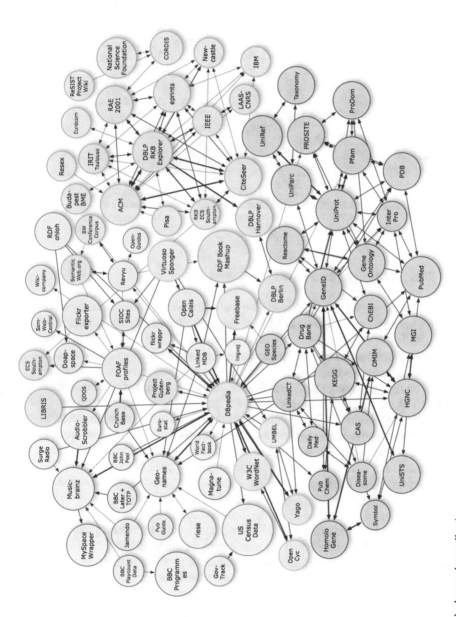

**Fig. 1.9** Linked open data collection

**Fig. 1.10**  Location-based services

### 1.4.4  Road Networks and Location-Based Services

Along with the advanced telecommunication technologies, location-based services have been a hot spot business area [21]. In our daily life, the road networks data with GPS on mobile devices are commonly used to guide us to our destinations where the common shortest path algorithm is used to specify our route (Fig. 1.10). In such services, high accuracy and fast response time are crucially required when running the algorithms on the underlying graph data. Because of the constraint of mobile devices (memory, CPU, network, etc.), the requirement of the algorithms is higher than that of off-line computing on cluster. On top of the traditional usage of the graph data on GPS that has relatively static addresses (nodes) and roads (edges), there are increasing requirements for dynamic services due to the increasing traffic congestion in the cities all around the world. Therefore, with the emerging trend of smart cities [22], new services are being developed to monitor and combine the traffic condition in order to provide real time route suggestions that can lead to the destinations within the shortest period of time. In addition, applications like *Foursquare*[17] is designed as a location-based services that provides recommendations of nearest places like restaurants or shopping malls based on user review or check-in data. In practice, combining the user information, from social network profiles, with geographical graph data helps the applications to provide more reliable recommendation [23].

---

[17]https://foursquare.com/.

## 1.4.5   Chemical and Biological Networks

The graph structure is also widely used in the research of fundamental science like chemical and biology. Chemical data is modeled as a graph by assigning atoms as nodes and bonds the edges between them [24]. Biological data is represented the same way, only with amino acids as the nodes and links between them as the edges (Fig. 1.11). In these domains, graph data is important for various operations as drug discovery and analysis. For example, in a protein–protein network, proteins with similar roles tend to serve similar metabolic functions. Thus, if we know the function of one protein, we can predict that all other proteins having a similar role would also have similar function [25]. In addition, pattern recognition techniques are used to find frequent subgraphs of a given graph. A prime example is the *Bio4j* project[18] that performs protein related querying and management over aggregated data from various systems (e.g., *Uniprot KB, Gene Ontology, UniRef, RefSeq*).

## 1.5   Graph Databases

In recent years, Graph Database Management Systems have been gaining popularity [26]. These systems are designed to efficiently store, index, traverse, and query graph databases. For example, *Neo4j*[19] is a popular and commercially supported open-source graph database which is implemented in Java [27]. It has been implemented as a disk-based transactional graph database. It provides efficient graph traversal operations, full ACID transaction support and convenient REST server interfaces. It also provides its own query language, called *Cypher*,[20] which can handle different kinds of queries. Cypher is a declarative query language similar to SQL. It allows its users to describe what should be selected, inserted, updated, or deleted from a graph database.

*Sparksee*[21] (formerly known as DEX) is bitmaps-based and disk-based graph database model written in C++. It is based on a labeled attributed multigraph, where all vertices and edges can have one or more attributes (attributed), edges have labels (labeled) and can be bidirectional (multigraph) [28]. In Sparksee, each graph is stored in a single file; values and resource identifiers are mapped by mapping functions; and maps are modeled as B+-tree. Sparksee uses key-value maps to provide the index of full data access to complement the bitmaps. Because the adjacency is usually sparse, the compressed bitmaps will use the space more efficiently. Queries are implemented as a combination of low level graph-oriented operations, which are highly optimized

---

[18]http://bio4j.com/.
[19]http://neo4j.com/.
[20]http://neo4j.com/developer/cypher-query-language/.
[21]http://www.sparsity-technologies.com/.

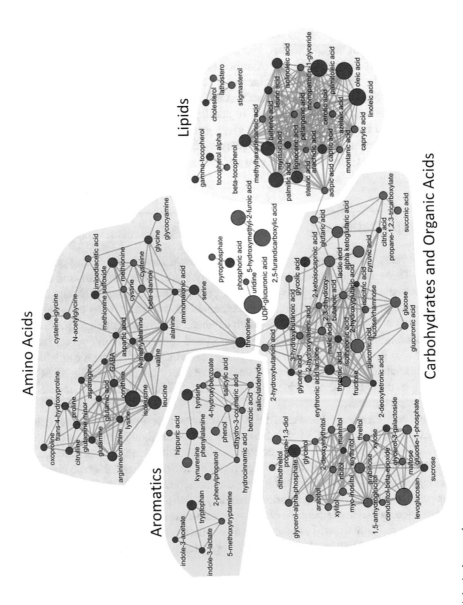

**Fig. 1.11** Biological networks

to get the maximum performance from the data structures. *HyperGraphDB*[22] is another open-source database focused on supporting generalized hypergraphs [29]. Data is stored in the form of key-value pairs on top of BerkeleyDB, and is indexed using B-trees. Hypergraphs differ from normal graphs in their ability for edges to point to other edges. This representation is useful in the modeling of graph data for artificial intelligence, bio-informatics, and other knowledge representations. Hypergraph supports online querying with a Java API. *InfiniteGraph*[23] is a distributed graph database system that supports efficient traversal of graphs across distributed data stores. This works by creating a federation of compute nodes operated through their Java API. *Titan*[24] is another cluster-based graph database that is designed for storing and querying graphs with hundreds of billions of vertices and edges. The main focus of Titan is compact graph serialization, graph data modeling, and the efficient execution of small, concurrent graph queries.

## 1.6   Does Hadoop Work Well for Big Graph Processing?

In principle, general-purpose distributed data processing frameworks such as Hadoop are well suited for analyzing unstructured and tabular data. However, such frameworks are not efficient for directly implementing iterative graph algorithms which often require multiple stages of complex joins [9]. In addition, the general-purpose join and aggregation mechanisms defined in such distributed frameworks are not designed to leverage the common patterns and structure in iterative graph algorithms. Therefore, such disregard of the graph structure leads to huge network traffic and missed opportunities to fully leverage important graph-aware optimization.

In practice, a fundamental aspect in the design of the Hadoop framework is that it requires the results of each single map or reduce task to be *materialized* into a local file on the Hadoop Distributed File System (HDFS), a distributed filesystem to store data files across multiple machines, before it can be processed by the following tasks (Fig. 1.12). This materialization step supports the implementation of a simple and elegant checkpointing/restarting fault-tolerance mechanism. However, from the graph-processing point of view, this materialization step dramatically hurts the performance of graph-processing algorithms (e.g., PageRank, Connected Component, Triangle Count) which are iterative in nature and typically aim to traverse the graph in a specific way. Hence, in practice, the efficiency of graph computations depends heavily on inter-processor bandwidth as graph structures are sent over the network iteration after iteration. In addition, the MapReduce framework does not directly support iterative data-analysis applications. To implement iterative programs, programmers might manually issue multiple MapReduce jobs and orchestrate

---

[22]http://www.hypergraphdb.org/.
[23]http://www.objectivity.com/products/infinitegraph/.
[24]http://thinkaurelius.github.io/titan/.

Iterations

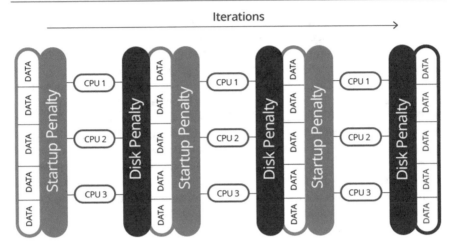

**Fig. 1.12**  MapReduce iteration

their execution with a driver program. In practice, the manual orchestration of an iterative program in MapReduce has two key problems:

- While much data might be unchanged from iteration to iteration, the data must be reloaded and reprocessed at each iteration, wasting I/O, network bandwidth, and processor resources.
- The termination condition might involve the detection of when a fix point is reached. The condition itself might require an extra MapReduce job on each iteration, again increasing resource use in terms of scheduling extra tasks, reading extra data from disk, and moving data across the network.

Several approaches have proposed Hadoop extensions (e.g., *HaLoop* [30], *Twister* [31], *iMapReduce* [32]) to optimize the iterative support of the MapReduce framework. For instance, the *HaLoop* system [30] has been designed to support iterative processing on the MapReduce framework by extending the basic MapReduce framework with two main functionalities: Caching the invariant data in the first iteration and then reusing them in later iterations in addition to caching the reducer outputs, which makes checking for a fix point more efficient, without an extra MapReduce job. The *iMapReduce* framework [32] supports the feature of iterative processing by keeping alive each map and reduce task during the whole iterative process. In particular, when all of the input data of a persistent task are parsed and processed, the task becomes dormant, waiting for the new updated input data. For a map task, it waits for the results from the reduce tasks and is activated to work on the new input records when the required data from the reduce tasks arrive. For the reduce tasks, they wait for the map tasks' output and are activated synchronously as in MapReduce. *Twister*[25] is a MapReduce runtime with an extended programming model that supports iterative MapReduce computations efficiently [31]. It uses a publish/subscribe messaging infrastructure for communication and data transfers, and supports long running map/reduce tasks. In particular, it provides programming extensions to MapReduce with broadcast and scatter type data transfers. Microsoft has also developed a project that provides an iterative MapReduce runtime for Windows Azure called *Daytona*.[26] In general, these approaches remain inefficient for the graph-processing case because the efficiency of graph computations depends heavily on inter-processor bandwidth as graph structures are sent over the network after each iteration.

Other approaches have attempted to implement graph-processing operations on top of the MapReduce framework (e.g., *Surfer* [33], *PEGASUS* [34]). For example, the *Surfer* system [33] has been presented as a large-scale graph-processing engine which is designed to provide two basic primitives for programmers: `MapReduce` and `propagation`. In this engine, MapReduce processes different key-value pairs in parallel, and propagation is an iterative computational pattern that transfers information along the edges from a vertex to its neighbors in the graph. In particular, to use the graph propagation feature in the Surfer system, the user needs to define two functions: `transfer` and `combine`. The `transfer` function is responsible for exporting the information from a vertex to its neighbors, while the `combine` function is responsible for aggregating the received information at each vertex. In addition, the Surfer system adopts a graph-partitioning strategy that attempts to divide the large graph into many partitions of similar sizes so that each machine can hold a number of graph partitions and manage the propagation process locally before exchanging messages and communicating with other machines. As a result, the propagation process can exploit the locality of graph partitions for minimizing the network traffic. *GBASE*[27] is another MapReduce-based system that uses a graph storage method, called *block compression*, which first partitions the input graph into a number of blocks [35]. According to the partition results, GBASE reshuffles the nodes so that the nodes belonging to the same partition are placed near to each other after which it compresses all nonempty block through a standard compression mechanism such as *GZip*.[28] Finally, it stores the compressed blocks together with some meta information into the graph storage. GBASE supports different types of graph queries including *neighborhood, induced subgraph, egonet, K-core*, and *cross-edges*. To achieve this goal, GBASE applies a grid selection strategy to minimize disk accesses and answer queries by applying a MapReduce-based algorithm that supports incidence matrix-based queries. Finally, *PEGASUS*[29] is a large-scale graph mining library that has been implemented on top of the Hadoop framework and sup-

---

[25]http://www.iterativemapreduce.org/.
[26]http://research.microsoft.com/en-us/projects/daytona/.
[27]http://systemg.research.ibm.com/analytics-search-gbase.html.
[28]http://www.gzip.org/.
[29]http://www.cs.cmu.edu/~pegasus/.

ports performing typical graph-mining tasks such as *computing the diameter of the graph*, *computing the radius of each node*, and *finding the connected components* using Generalized Iterative Matrix-Vector multiplication (GIM-V) which represents a generalization of normal matrix-vector multiplication [34, 36]. The library has been utilized for implementing a MapReduce-based algorithm for discovering patterns on near-cliques and triangles on large-scale graphs [37]. In practice, GBASE and PEGASUS are unlikely to be intuitive for most developers, who might find it challenging to think of graph processing in terms of matrices. Also, each iteration is scheduled as a separate Hadoop job with increased workload: When the graph structure is read from disk, the map output is spilled to disk and the intermediate result is written to the HDFS.

## 1.7   BSP Programming Model and Google Pregel

Bulk Synchronous Parallel (BSP) is a parallel programming model that uses a message passing interface (MPI) to address the scalability challenge of parallelizing jobs across multiple nodes [38]. In principle, BSP is a vertex-centric programming model where the computations on vertices are represented as a sequence of *super-steps* (iterations) with synchronization between the nodes participating at superstep barriers. In particular, in each iteration, every vertex that is involved in computation, (1) receives its neighbors updated values from previous iteration, (2) the vertex then will be updated by received values, (3) sends its updated value to its adjacent vertices that will be available to them in the next superstep. At each iteration (superstep), each vertex can vote to be active or inactive in the next iteration (Fig. 1.13). Such a programming model can be seen as a graph extension of the actor programming model [39] where each vertex represents an actor and edges represent the communication channel between actors. In such model, users can focus on specifying the computation on the graph nodes and the communication among them without worrying about the specifics of the underlying organization or resource allocation of the graph data.

In 2010, the *Pregel* system [40], presented by Google and implemented in C/C++, has been introduced as the first BSP implementations that provides a native API specifically for programming graph algorithms using a *"think like a vertex"* computing paradigm where each vertex has a value and can exchange message with other graph vertices in a number of iterations. During each iteration of the computation, each vertex can receives messages, updates its value and sends messages to its neighbor vertices (Fig. 1.14). In order to avoid communication overheads, Pregel preserves data locality by ensuring computation is performed on locally stored data. In particular, Pregel distributes the graph vertices to the different machines of the cluster where each vertex and its associated set of neighbors are assigned to the same node. Graph-processing algorithms are then represented as supersteps where each step defines what each participating vertex has to compute and edges between vertices represent communication channels for transmitting computation results from

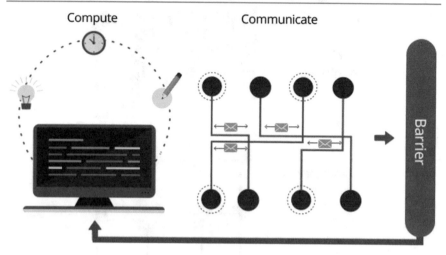

Compute Communicate

Barrier

**Fig. 1.13** BSP programming model

Input Message (s) — VertexData — EdgeData — Output Message (s)

EdgeData

**Fig. 1.14** Vertex-based programming model

one vertex to another. In particular, at each superstep, a vertex can execute a user-defined function, send or receive messages to its neighbours (or any other vertex with a known ID), and change its state from active to inactive. Pregel supersteps are synchronous. That means each superstep is concluded once all active vertices of this steps have completed their computations and all of the exchanged messages among the graph vertices have been delivered. Pregel can start the execution of a new superstep $(S + 1)$ only after the current superstep $(S)$ completes its execution. Each superstep ends with a waiting phase, synchronization barrier (Fig. 1.15), that ensures that messages sent from one superstep are correctly delivered to the subsequent step. In each superstep, a vertex may vote to halt (inactive status) if it does not receive any message and it can also be re-activated once it receives a message at any subsequent superstep (Fig. 1.16). Thus, in each superstep, only the active vertices are involved in the computation process results in significant reduction in the communication overhead (Fig. 1.17), a main advantage against the Hadoop-based processing for graphs. The whole graph- processing operation terminates when all vertices are inactive and no more messages are in transit between the vertices of the graph. In Pregel, the input graph is loaded once at the beginning of the program and all computations are executed in-memory. Pregel uses a master/workers model where the master node

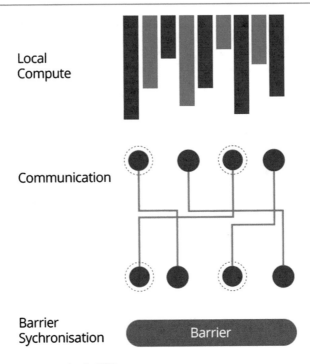

Local
Compute

Communication

Barrier
Sychronisation

**Fig. 1.15**  Supersteps execution in BSP

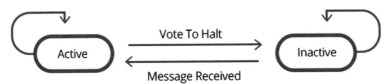

Vote To Halt

Active                    Inactive

Message Received

**Fig. 1.16**  Vertex voting in BSP

is responsible for coordinating synchronization at the superstep barriers while each worker independently invokes and executes the `compute()` function on the vertices of its assigned portion of the graph and maintains the message queue to receive messages from the vertices of other workers.

## 1.8  Pregel Extensions

The introduction of Google's Pregel has triggered much interest in the field of large-scale graph data processing and inspired the development of several Pregel-based systems which attempt to exploit different optimization opportunities. For instance,

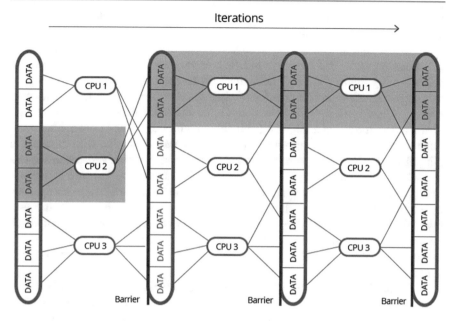

**Fig. 1.17** BSP execution for iterative processing

$GPS^{30}$ is another open-source Java implementation of Google's Pregel which comes from Stanford InfoLab [41]. GPS extends the Pregel API to allow certain global computation tasks to be specified and run by a master worker. In particular, it provides an additional function, `master.compute()`, that provides access to all of the global aggregated values, and store the global values which are transparent to the vertices. The global aggregated values can be updated before they are broadcast to the workers. GPS also offers the Large Adjacency List Partitioning (LALP) mechanism as an optional performance optimization for algorithms that send the same message to all of its neighbours. In particular, LALP works by partitioning the adjacency lists of high-degree vertices across different workers. For each partition of the adjacency list of a high-degree vertex, a mirror of the vertex is created in the worker that keeps the partition. When a high-degree vertex broadcasts a message to its neighbors, at most one message is sent to its mirror at each machine. Then, the message is forwarded to all its neighbors in the partition of the adjacency list of the high-degree vertex. This mechanism works well for algorithms like PageRank, weakly connected components (WCC), and single source shortest path (SSSP) with unit edge weights but does not work well for some other algorithms like distributed minimal spanning tree construction (DMST). Furthermore, GPS applies a dynamic repartitioning strategy based on the graph-processing workload in order to balance the workload among all workers and reduce the number of exchanged messages over

---

[30]http://infolab.stanford.edu/gps/.

the network. In particular, GPS exchanges vertices between workers based on the amount of data sent by each vertex.

Similar to GPS, *Mizan* is an open-source project developed in C++ by KAUST, in collaboration with IBM Research [42, 43]. Mizan proposed a runtime-based load balancing algorithm for graph computations, through dynamic repartitioning, to maximize the end-to-end runtime for graph systems. Mizan's dynamic repartitioning strategy is based on monitoring the runtime characteristics of the graph vertices (e.g., their execution time, and incoming and outgoing messages) and uses this information, at the end of every superstep, to construct a migration plan with the aims of minimizing the variations across workers by identifying which vertices to migrate and where to migrate them to. The migration plan is executed between supersteps transparently to avoid interfering with the user's computations. Mizan accepts any type of graph-partitioning technique because, unlike most graph systems, it uses a distributed hash table (DHT) to maintain the location of vertices. The DHT is updated whenever a vertex moves between different workers as part of the migration plan to ensure that the messages in the next superstep can reach the new location of the vertex.

*Pregelix*[31] is a large-scale graph-processing platform that applies set-oriented, iterative dataflow approach to implement the BSP-based Pregel programming model [44]. In particular, Pregelix treats the messages and vertex states in the graph computation as relational tuples with a well-defined schema and uses relational database-style query evaluation techniques to execute the graph computation. For example, Pregelix treats message exchange as a join operation followed by a group-by operation that embeds functions which capture the semantics of the graph computation program. Therefore, Pregelix generates a set of alternative physical evaluation strategies for each graph computation program and uses a cost model to select the target execution plan among them. The execution engine of Pregelix is *Hyracks* [45], a general-purpose shared-nothing dataflow engine. Given a graph-processing job, Pregelix first loads the input graph dataset (the initial Vertex relation) from a distributed file system, i.e., HDFS, into a Hyracks cluster and partitions it using a user-defined partitioning function across the worker machines. Pregelix leverages B-tree index structures from the Hyracks storage library to store partitions of Vertex on worker machines. During the supersteps, at each worker node, one (or more) local indexes are used to store one (or more) partitions of the Vertex relation. After the eventual completion of the overall graph computation, the partitioned Vertex relation is scanned and dumped back to HDFS.

*Pregel+*[32] is another Pregel-based project implemented in C/C++ with the aim of reducing the number of exchanged messages between the worker nodes using a mirroring mechanism. In particular, Pregel+ selects the vertices for mirroring based on a cost model that analyzes the tradeoff between mirroring and message combining. *Giraph++* [46] has proposed a *"think like a graph"* programming paradigm that

---

[31]http://pregelix.ics.uci.edu/.
[32]http://www.cse.cuhk.edu.hk/pregelplus/.

opens the partition structure to the users so that it can be utilized within a partition in order to bypass the heavy message passing or scheduling facilities. In particular, the graph-centric model can make use of off-the-shelf sequential graph algorithms in distributed computation, thus allowing asynchronous computation to accelerate convergence rates and naturally support existing partition-aware parallel/distributed algorithms.

## 1.9   Giraph: BSP + Hadoop for Graph Processing

In 2012, *Apache Giraph* has been introduced as an open-source project that clones the ideas and implementation of Pregel specification in Java on top of the infrastructure of the Hadoop framework (Fig. 1.18). In principle, the relationship between the proprietary Pregel system and Giraph Apache open-source project is similar to the relationship between the Google MapReduce framework and the Apache Hadoop project (Fig. 1.19). Giraph has been initially implemented by Yahoo!. It had a growing community of users and developers worldwide so that Giraph has become a popular graph-processing framework. For example, Facebook built its Graph Search services using Giraph. Currently, Giraph enlists contributors from Yahoo!, Facebook, Twitter and Linkedin. Giraph runs graph-processing jobs as map-only jobs on Hadoop and uses HDFS for data input and output. It uses a set of machines (workers) to process large graph datasets. One of the machines plays the role of master to coordinate with other slave workers. The master is also responsible for global synchronization, error handling, and assigning partitions to workers. Giraph added many features beyond the basic Pregel including sharded aggregators, out-of-core computation, master computation, edge-oriented input, and more. Inherited from Hadoop's Architecture, Giraph uses *Apache ZooKeeper*[33] for maintaining configuration information, distributed synchronization, checkpointing, and failure recovery schemes. ZooKeeper supports high availability via using redundant services where clients can ask another ZooKeeper leader if its first call has failed (Fig. 1.20).

Primarily, Giraph provides a distributed execution engine for graph computations. In addition, it supports a programming model that allows its user to focus on designing the logic of their graph computation without the need to worry about how the underlying graph is stored in disk or loaded in memory or how the execution of the computation is distributed among the nodes of the computing cluster with fault-tolerance considerations. In Giraph, each vertex of the graph is identified by a unique ID. Each vertex also has other information such as a vertex value, a set of edges with an edge value for each edge, and a set of messages sent to it. Graph-processing programs are expressed as a sequence of iterations (supersteps). During a superstep, the framework starts a user-defined function for each vertex, conceptually in parallel. The user-defined function specifies the behavior at a single vertex $V$ and a single

---

[33]http://zookeeper.apache.org/.

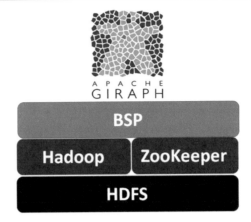

**Fig. 1.18** Layered architecture of Giraph

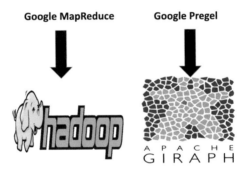

**Fig. 1.19** Pregel and Giraph

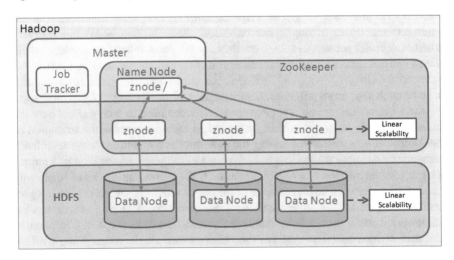

**Fig. 1.20** Zookeeper synchronization

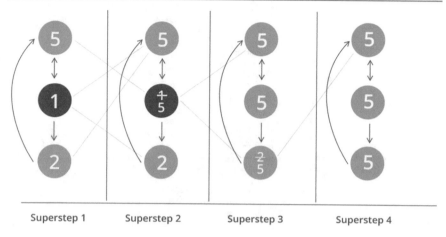

Superstep 1          Superstep 2          Superstep 3          Superstep 4

**Fig. 1.21** BSP example

superstep $S$. The function can read messages that are sent to $V$ in superstep $S - 1$, send messages to other vertices that are received at superstep $S + 1$, and modify the state of $V$ and its outgoing edges. Messages are typically sent along outgoing edges, but you can send a message to any vertex with a known identifier. Each superstep represents atomic units of parallel computation.

Figure 1.21 illustrates an example for the communicated messages between a set of graph vertices for computing the maximum vertex value. In this example, in Superstep 1, each vertex sends its value to its neighbor vertex. In Superstep 2, each vertex compares its value with the received value from its neighbor vertex. If the received value is higher than the vertex value, then it updates its value with the higher value and sends the new value to its neighbor vertex. If the received value is lower than the vertex value, then the vertex keeps its current value and votes to halt. Hence, in Superstep 2, only the vertex with value 1 updates its value to higher received value (5) and sends its new value. This process happens again in Superstep 3 for the vertex with the value 2, while in Superstep 4 all vertices vote to halt and the program ends.

Similar to the Hadoop framework, Giraph is an efficient, scalable, and fault-tolerant implementation on clusters of thousands of commodity computers, with the distribution-related details hidden behind an abstraction. On a machine that performs computation, it keeps vertices and edges in memory and uses network transfers only for messages. The model is well suited for distributed implementations because it does not show any mechanism for detecting the order of execution within a super-step and all communications are from superstep $S$ to superstep $S + 1$. During program execution, graph vertices are partitioned and assigned to workers. The default partition mechanism is hash-partitioning, but custom partition is also supported. Giraph applies a master/worker architecture, illustrated in Fig. 1.22. The master node assigns partitions to workers, coordinates synchronization, requests checkpoints, and

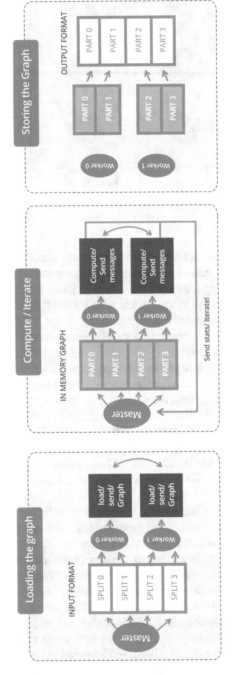

**Fig. 1.22** Execution flow of Giraph

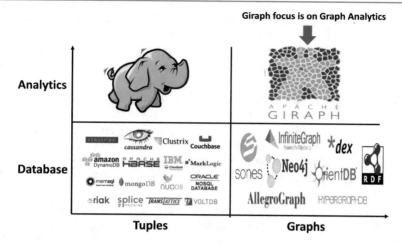

**Fig. 1.23** Giraph versus graph databases

collects health statuses. Workers are responsible for vertices. A worker starts the `compute()` function for the active vertices. It also sends, receives, and assigns messages with other vertices. During execution, if a worker receives input that is not for its vertices, it passes it along.

We would like to bring the attention of the reader to differentiate between large-scale graph-processing and analytics systems (e.g., Girpah) and other types of specialized graph database systems (e.g., Neo4J, Sparsity, InfiniteGraph, Hyper-graphDB) which has been introduced in Sect. 1.5. Giraph is not a database. For example, it does not handle updates or deletes on the graph. However, it is a distributed graph-processing system. In principle, the main focus of systems like Giraph is on batch processing of analytics workloads with off-line, long running, and expensive graph computations while graph database systems are mainly focusing on serving online transactional workloads with interactive and low-latency requirements. To illustrate, the difference in the usage scenario between Giraph and graph database systems is analog to the difference in the usage scenario between the Hadoop framework and traditional relational database systems (e.g., MySQL, Postgres, Oracle, SQL Server) or NoSQL databases (e.g., HBase, Cassandra) (Fig. 1.23).

## 1.10  Book Roadmap

Chapter 2 takes on Giraph as a platform. Keeping in view that Giraph uses Hadoop as its underlying execution engine, we explain how to set up Hadoop in different modes, how to monitor it, and how to run Giraph on top of it using its binaries or source code. We then move to explaining, how to use Giraph. We start by running an example job in different Hadoop modes and then approach more advanced topics like monitoring Giraph application life cycle and monitoring Giraph jobs using different methods. Giraph is a very flexible platform and its behavior can be tuned in many ways. We explain the different methods of configuring Giraph and end the chapter by giving a detailed description of setting up a Giraph project in Eclipse and IntelliJ IDE.

Chapter 3 provides an introduction to Giraph programming. We introduce the basic Giraph graph model and explain how to write a Giraph program using the vertex similarity algorithm as a use case. We explain three different ways of writing the driver program and their pros and cons. For loading data into Giraph, it comes packaged with numerous input formats for reading different formats of data. We describe each of the formats with examples and end the chapter with the description of Giraph output formats.

Chapter 4 discusses the implementation of some popular graph algorithms including PageRank, connected components, shortest paths, and triangle closing. In each of these algorithms, we give an introductory description and show some of its possible applications. Then using a sample data graph, we show how the algorithm works. Finally, we describe the implementation details of the algorithm in Giraph.

Chapter 5 spots the light on advanced Giraph programming. We start by discussing common Giraph algorithmic optimizations and how those optimizations may improve the performance and flexibility of the algorithms implemented in Chap. 4. We explain different graph optimizations to enable users to implement complex graph algorithms. Then, we discuss a set of tunable Giraph configurations that controls Giraph's utilization of the underlying resources. We also discuss how to change Giraph's default partitioning algorithm and how to write a custom graph input and output format. We then talk about common Giraph runtime errors and finalize the chapter with information on Giraph's failure recovery.

Following the introduction of Google's Pregel and Apache Giraph, the field of graph analytics systems has attracted a lot of interest. Several systems have been introduced to tackle this challenge. In Chap. 6, we highlight two of these systems, GraphX and GraphLab. We describe their program abstractions and their programming models. We also highlight the main commonalities and differences between these systems and Apache Giraph.

## 1.11   How to Use the Code Examples of This Book?

The code examples of this book can be found on the Github repository[34] of the book. The repository folders are structured according to the chapters of the book. To download the examples on your computer, make sure that you have git[35] installed.

---

[34]https://github.com/sakrsherif/GiraphBookSourceCodes.
[35]https://git-scm.com/downloads.

At the terminal, go to the directory where you would like to download the source codes and type the following command:

```
git clone https://github.com/sakrsherif/
    GiraphBookSourceCodes
```

This command will ask you for your Github username and password. If you do not have an account already, you can create one free of charge. Alternatively, you can go to the repository page of the book and click on **Clone or download** button displayed in green color. It will give you an option to **Download ZIP** which you can extract to a directory on your computer after download.

# Installing and Getting Giraph Ready to Use

**2**

## 2.1 Installing Hadoop

Apache Giraph runs on top of Apache Hadoop which has three different installation modes:

1. *Standalone*: In this mode, Hadoop runs as a single Java process on a single node. This mode is useful for debugging.
2. *Pseudo-distributed*: In this mode, Hadoop runs on a single node but each Hadoop daemon runs in a separate Java process.
3. *Fully Distributed*: In this mode, Hadoop runs on a cluster of multiple nodes in a fully distributed fashion.

If Hadoop is already installed and $HADOOP_HOME environment variable points to the Hadoop installation, Giraph can execute its code transparently. In the following sections, we will explain the installation procedure of Hadoop in all of the three modes and later on, we will explain how to deploy Giraph on an existing Hadoop installation. Each section will follow the following steps to successfully install and run Hadoop:

1. Hadoop installation.
2. Hadoop installation verification by running an example Hadoop job.

The node on which we will be installing Hadoop and later on Giraph, has the following settings:

```
OS: Linux (Ubuntu preferable)
Admin account: hadoopAdmin
Hostname: hadoopNode01
```

We will be using Apache Hadoop 1.2.1 and Apache Giraph 1.1.0.

© Springer International Publishing AG 2016
S. Sakr et al., *Large-Scale Graph Processing Using Apache Giraph*,
DOI 10.1007/978-3-319-47431-1_2

## 2.1.1   Single-Node Local Mode Installation

1. Install Java and verify it by checking its version:

```
sudo apt-get install openjdk-7-jdk
java -version
```

   This should display the current version of Java installed.
2. Create a user `hadoopUser` under group `hadoopGroup`. This user will be used to install and run Hadoop.

```
sudo groupadd hadoopGroup
sudo useradd -g hadoopGroup -m hadoopUser
sudo passwd hadoopUser
sudo chsh -s /bin/bash hadoopUser #change shell
    to bash
```

3. Switch to `hadoopAdmin` and download Hadoop.

```
su - hadoopAdmin
cd /usr/local
sudo wget
https://archive.apache.org/dist/hadoop/common/
    hadoop-1.2.1/hadoop-1.2.1.tar.gz
#http://archive.apache.org/dist/hadoop/core/
    hadoop-0.20.203.0/hadoop-0.20.203.0rc1.tar.gz
```

4. Unpack downloaded Hadoop file and rename the unpacked hadoop directory to a more readable name, i.e., `hadoop`.

```
sudo tar xzf hadoop-1.2.1.tar.gz
sudo mv hadoop-1.2.1.tar.gz hadoop
#sudo tar xzf hadoop-0.20.203.0rc1.tar.gz
#sudo mv hadoop-0.20.203.0 hadoop
```

5. Make `hadoopUser` the owner of the `hadoop` directory and switch back to `hadoopUser`. Changing the ownership of the `hadoop` installation directory will allow us to modify the configuration files using `hadoopUser`.

```
sudo chown -R hadoopUser:hadoopGroup hadoop
su hadoopUser
```

6. Create Hadoop and Java environment variables, and add Hadoop scripts to the `$PATH` variable by adding the following lines at the end of `/home/hadoop User/.bashrc` file and save it.

```
export HADOOP_HOME=/usr/local/hadoop
export JAVA_HOME=/usr/lib/jvm/java-7-openjdk-
    amd64
export PATH=$PATH:$HADOOP_HOME/bin
```

If you have Java installed in a different directory, replace the value of JAVA_HOME with its path.

7. Reload the .bashrc file to create the new environment variables.

```
source ~/.bashrc
```

8. Modify the $HADOOP_HOME/conf/hadoop-env.sh file by adding/updating the following lines.

```
export JAVA_HOME=/usr/lib/jvm/java-7-openjdk-
    amd64
export HADOOP_OPTS=-Djava.net.preferIPv4Stack=
    true
```

### Installation Verification

9. We can verify the Hadoop installation by running a MapReduce job. We will do it by running the wordcount example job that comes packaged with Hadoop. For this, we first download a book from http://www.gutenberg.org in the /tmp/gutenberg directory using the following command:

```
mkdir -p /tmp/gutenberg
wget http://www.gutenberg.org/files/5000/5000-8.
    txt -P /tmp/gutenberg/
```

10. Run the wordcount example:

```
cd $HADOOP_HOME
hadoop jar hadoop*examples*.jar wordcount /tmp/
    gutenberg /tmp/gutenberg-output
```

11. Verify that the output files have been created successfully.

```
ls /tmp/gutenberg-output
```

The above command should return with following files/directories:

```
part-r-00000   _SUCCESS
```

The _SUCCESS is an empty file indicating the successful execution of the job. part-r-00000 file contains the actual output of the wordcount job. The "r" in the output file name shows that this file has been produced by a *Reducer* task. The "00000" is the ID of the file which increments with the addition of each reducer. If there are more than one reducers executing the MapReduce job, there will be one output file for each reducer task.

12. Display the output of the wordcount job:

```
less /tmp/gutenberg-output/part-r-00000
```

The above command produces a two-column output similar to the following listing, where the first column is the word and the second column is the count of the word in the input file:

```
"(Lo)cra"           1
"1490    1
"1498,"1
"35"    1
"40,"   1
"AS-IS".            1
```

13. If you run the wordcount example again, you will get the following exception:

```
org.apache.hadoop.mapred.
    FileAlreadyExistsException:  Output  directory
    /tmp/gutenberg-output  already  exists
```

This is because Hadoop does not overwrite an existing directory. You should either delete this directory or specify a new output directory.

14. Single-node Hadoop installation in Local Mode is now complete. If you want to run Giraph in this mode, you can jump to Sect. 2.3 for installing Giraph.

## 2.1.2  Single-Node Pseudo-Distributed Installation

For installing Hadoop in a pseudo-distributed mode, follow the steps until step 8 in Sect. 2.1.1 and then continue with the following steps:

1. Switch to hadoopAdmin and create directories for Hadoop to store its temporary files, files of the NameNode and the DataNode.

```
su - hadoopAdmin
sudo mkdir -p /app/hadoop/tmp
sudo mkdir -p /app/hadoop/data/namenode
sudo mkdir -p /app/hadoop/data/datanode
```

2. Make hadoopUser the owner of the temporary, NameNode, and DataNode directories and change their permissions such that hadoopUser has full access and the rest can read and execute.

```
sudo chown -R hadoopUser:hadoopGroup /app/hadoop
    /tmp
sudo chmod -R 755 /app/hadoop
```

3. Make sure that the machine has been assigned the hostnames hadoopNode01 and localhost. Edit the /etc/hosts file and add/update the following lines. <ip-address> in the following should be replaced by the ip address of the machine on which Hadoop is being installed.

```
127.0.0.1          localhost
<ip-address>       hadoopNode01
```

For single-node installations, only `localhost` can be used for all communication but it is better to use `hadoopNode01` because other machines can talk
to this machine using this hostname. This is helpful in multi-node installation. If
the machine you are installing Hadoop on, has a dynamic IP (which is not fixed),
instead of using a separate ip for `hadoopNode01`, you can use `127.0.0.1`.
For this scenario, modify the `/etc/hosts` file as follows:

```
127.0.0.1          localhost hadoopNode01
```

4. Switch to `hadoopUser` using `su hadoopUser` command and edit the
   `core-site.xml` file in `$HADOOP_HOME/conf` directory and make sure
   that the following four properties are added to the file between
   `<configuration>` tags.

```
<property>
    <name>hadoop.tmp.dir</name>
    <value>/app/hadoop/tmp</value>
</property>
<property>
    <name>fs.default.name</name>
    <value>hdfs://hadoopNode01:54310</value>
</property>
```

The first property tells Hadoop where to store its temporary files, whereas the
second property tells hadoop the address of the HDFS master node which in our
case is the machine itself.

5. Edit `mapred-site.xml` in `$HADOOP_HOME/conf` directory and add the
   following properties.

```
<property>
    <name>mapred.job.tracker</name>
    <value>hadoopNode01:54311</value>
</property>

<property>
    <name>mapred.tasktracker.map.tasks.maximum</
        name>
    <value>4</value>
</property>

<property>
    <name>mapred.map.tasks</name>
    <value>4</value>
</property>
```

The first property tells Hadoop on which machine and port to start the Job Tracker service. The second property is used by each Task Tracker to know the total number of Map tasks it can launch. The third property tells Hadoop how many Map tasks should be launched per job. Hadoop defaults for both the properties is 2 but some of Giraph unit tests fail with this default value. We will use these settings for standalone and pseudo-distributed installations.

6. Add the following properties between the `<configuration>` tags in hdfs-site.xml in $HADOOP_HOME/conf directory.

```
<property>
    <name>dfs.name.dir</name>
    <value>/app/hadoop/data/namenode</value>
</property>
<property>
    <name>dfs.data.dir</name>
    <value>/app/hadoop/data/datanode</value>
</property>
<property>
    <name>dfs.replication</name>
    <value>1</value>
</property>
```

The first and second properties tell Hadoop where to store the data of the namenode and the datanode respectively. Keep in mind that the data stored in HDFS is actually stored in dfs.data.dir so make sure that it has enough space to hold the data. The third property tells HDFS to store only one copy of the data stored on it. Hadoop's default for this property to ensure fault tolerance is 3 but as we are using only one machine, one copy of the data is enough. If the machine fails, the data will not be accessible anyways, because all the replicas are stored on its disk.

7. When Hadoop is started, it logs in to all slave nodes to start Hadoop services. You should enable password-less SSH for all slave machines otherwise the machines will ask for a password for each SSH access Hadoop makes. Use the following commands to enable password-less SSH. Press the enter key for the question about the file for saving the key to accept the default file.

```
ssh-keygen -t rsa -P ""
cat $HOME/.ssh/id_rsa.pub >> $HOME/.ssh/
    authorized_keys
```

8. Now SSH to the machine from the machine itself.

```
ssh hadoopNode01
```

If this is the first time you are doing an SSH, the machine will ask for a password. After you enter the password, it will prompt you if it should add the public RSA key into $HOME/.ssh/known_hosts. You should reply with a yes. In case you are unable to SSH and receive the message:

```
ssh: connect to host hadoopnode01 port 22:
    Connection refused
```

the SSH service might not be running. You can start the SSH service using the following command:

```
su - hadoopAdmin
sudo service ssh start
```

If the above command fails with a message:

```
ssh: unrecognized service
```

you might not have an SSH server installed. You should run the following commands to install the openssh-server:

```
sudo apt-get update
sudo apt-get upgrade
sudo apt-get install openssh-server
```

When you install the openssh-server, the SSH service automatically gets started, but to be sure, try to start the SSH service:

```
sudo service ssh start
```

The above command should return with a message that the Job is already running. Now try to SSH again using hadoopUser:

```
su hadoopUser
ssh hadoopNode01
```

Now you should be able to SSH successfully.

9. A Hadoop setup consists of a master and many slaves. As we are using a single node, the same machine will be acting as the master and the slave. Add the following line to the files $HADOOP_HOME/conf/masters and $HADOOP_HOME/conf/slaves.

```
hadoopNode01
```

10. Initialize the HDFS by formatting it

```
hadoop namenode -format
```

11. Start HDFS and MapReduce services using the following commands:

```
start-dfs.sh
start-mapred.sh
```

12. Verify that all the Hadoop services processes are running on the machine by typing the following command:

```
jps
```

Make sure that the output of the above command contains all of the following five processes:

```
NameNode
SecondaryNameNode
DataNode
JobTracker
TaskTracker
```

The first three processes belong to HDFS, whereas the last two processes belong to MapReduce. If any of the processes is missing, you will have to debug it using the information from https://www.hadoop.apache.org/.

13. To stop Hadoop, you should first stop MapReduce and then HDFS. This will ensure that HDFS is not shutdown when a Hadoop job is in the middle of writing data to HDFS. We need Hadoop services to be up and running for the next steps so do not shut them down now.

```
stop-mapred.sh
stop-dfs.sh
```

**Installation Verification**

1. We can verify the Hadoop installation by running a MapReduce job. We will do it by running the wordcount example job that comes packaged with Hadoop. For this, we first download a book from http://www.gutenberg.org in the /tmp/gutenberg directory using the following command:

```
mkdir -p /tmp/gutenberg
wget http://www.gutenberg.org/files/5000/5000-8.
   txt -P /tmp/gutenberg/
```

2. Copy the downloaded book to HDFS:

```
hadoop dfs -mkdir /user/hadoopUser/gutenberg
hadoop dfs -copyFromLocal /tmp/gutenberg/5000-8.
   txt /user/hadoopUser/gutenberg
```

3. Verify that the book has been successfully copied to HDFS:

```
hadoop dfs -ls /user/hadoopUser/gutenberg
```

The above command will list the file name gutenberg with its size.

4. Run the `wordcount` example:

```
cd $HADOOP_HOME
hadoop jar hadoop*examples*.jar wordcount /user/
    hadoopUser/gutenberg /user/hadoopUser/
    gutenberg-output
```

5. Verify that the output file has been created successfully.

```
hadoop dfs -ls /user/hadoopUser/gutenberg-output
```

The above command should return with following files/directories:

```
-rw-r--r-- /user/hadoopUser/gutenberg-output/
    _SUCCESS
drwxr-xr-x /user/hadoopUser/gutenberg-output/
    _logs
-rw-r--r-- /user/hadoopUser/gutenberg-output/
    part-r-00000
```

The _SUCCESS is an empty file indicating the successful execution of the job. _logs folder contains the logs for the job. part-r-00000 file contains the actual output of the wordcount job.

6. Display the output of the wordcount job:

```
hadoop dfs -cat /user/hadoopUser/gutenberg-
    output/part-r-00000
```

It will produce a two-column output in which the first column is the word and the second column is the count of the word in the input file.

7. If the above steps have been completed successfully, single-node Hadoop installation in pseudo-distributed mode is now complete. If you want to run Giraph in this mode, you can jump to Sect. 2.3 for installing Giraph.

## 2.1.3 Multi-node Cluster Installation

Before installing Hadoop on a cluster, we recommend to install Hadoop in a pseudo-distributed mode first as explained in Sect. 2.1.2. When you have verified the pseudo-distributed installation, use the following steps to add another node as a slave. The resulting cluster will have three nodes, i.e., hadoopNode01, hadoopNode02, and hadoopNode03. hadoopNode01 will act as a Master whereas the other two nodes will act as slaves. You can use these steps to add as many nodes as you want.

### 2.1.3.1 Preparing a Slave Node

Use the following steps to prepare a new machine to be added as a slave to the Hadoop cluster. We assume the following settings for the slave node:

```
OS: Linux (Ubuntu preferable)
Admin account: hadoopAdmin
Hostname: hadoopNode02
```

1. From Sect. 2.1.1, follow step 1 to install Java and step 2 to create `hadoopUser` and `hadoopGroup`. `hadoopUser` will be used to run Hadoop in cluster mode.
2. Create Hadoop directories and assign permissions using steps 1 and 2 in Sect. 2.1.2. You do not need to create the `namenode` directory as the slave node will not be running the NameNode daemon.
3. Make sure that the slave machine `hadoopNode02` has network access to the Master machine (`hadoopNode01`) and vice-versa using the hostnames. Using the `hadoopAdmin` user, edit the `/etc/hosts` file and add/update the following lines. The text <ip-address-*> in the following should be replaced with the corresponding ip addresses.

```
<ip-address-master> hadoopNode01
<ip-address-current-node> hadoopNode02
```

Ping the Master node to check network access:

```
ping hadoopNode01
```

If the Master node is not accessible, it could either be a network or a security issue.
4. Follow step 8 in Sect. 2.1.2 to install/start SSH server by replacing `hadoopNode01` with `hadoopNode02`.
5. Copy the Hadoop directory from master to each slave node:

```
su hadoopAdmin
sudo rsync -a --progress hadoopUser@hadoopNode01
    :/usr/local/hadoop-1.2.1 /usr/local/
```

This will copy the Hadoop directory from the master node along with its permissions.
6. Repeat the above steps on `hadoopNode03` by replacing `hadoopNode02` in the above text to `hadoopNode03`.

### 2.1.3.2  Preparing the Master Node

In the pseudo-distributed installation we described previously, `hadoopNode01` was acting as a master as well as the slave node. In the clustered mode, we will use `hadoopNode01` only as a master. Use the following steps to configure `hadoopNode01` to act as a master in a multi-node Hadoop installation. Stop Hadoop using the script `stop-all.sh` before continuing with the following steps.

1. Hadoop treats all hostnames mentioned in the slaves file present in the conf directory inside $HADOOP_HOME directory. On hadoopNode01, switch to hadoopUser and modify the slaves file as follows:

```
hadoopNode02
hadoopNode03
```

2. Make sure that the ip addresses of the master and the slave nodes are entered in the /etc/hosts file as follows:

```
127.0.0.1            localhost
<ip-address>         hadoopNode01
<ip-address>         hadoopNode02
<ip-address>         hadoopNode03
```

Remember that unlike in pseudo-distributed mode, the host name hadoop Node01 should not be used for the ip 127.0.0.1 rather a corresponding network ip should be used. Otherwise Hadoop's daemons will listen to ports of 127.0.0.1 which means that the Hadoop daemons running on the remote nodes will not be able to connect to the daemons at the master node.

3. Make sure that the master node can access all slave nodes:

```
ping hadoopNode02
ping hadoopNode03
```

4. Confirm that hadoopUser can SSH to both the slave nodes.

```
ssh hadoopUser@hadoopNode02
exit
ssh hadoopUser@hadoopNode03
exit
```

If you are told that the authenticity of the slave node cannot be established and if you are sure you want to continue connecting, enter yes as an answer.

5. Enable password-less SSH from the master node to each slave node. Make sure you are switched to hadoopUser. If not, do so using the command "su hadoopUser."

```
ssh-copy-id -i $HOME/.ssh/id_rsa.pub
    hadoopUser@hadoopNode02
ssh-copy-id -i $HOME/.ssh/id_rsa.pub
    hadoopUser@hadoopNode03
```

The above commands will copy the public SSH key of the master to the slaves' .ssh/authorized_keys file in the home directory of hadoopUser. You will be asked for the password of hadoopUser for each slave node.

6. Follow step 4 to confirm that you can login to the slaves without providing a password.
7. Cleanup HDFS directories from the previous pseudo-distributed installation:

```
rm -r /app/hadoop/tmp/*
rm -r /app/hadoop/data/namenode/*
rm -r /app/hadoop/data/datanode/*
```

8. Follow steps 10–12 to format HDFS, start Hadoop and verify if it is running. To verify that the Hadoop installation is working properly, follow the steps in the "Installation Verification" subsection in Sect. 2.1.2.

## 2.2  Monitoring Hadoop

There are a number of ways in which Hadoop can be monitored. There are tools like Ganglia and Nagios which can be used for this purpose, but in this book we will introduce you to some basic monitoring techniques which will provide you with the foundation.

### 2.2.1  Hadoop Web User Interfaces

The easiest way to monitor hadoop is to use its Web UI. Hadoop has two web interfaces, one for HDFS and the other for MapReduce. These interfaces are served through the master node as both the HDFS NameNode and the JobTracker are running on the master node but it is not always the case.

#### 2.2.1.1  HDFS Web UI

The HDFS Web UI can be found at the URL http://www.<master>:50070 which in our case is http://www.hadoopnode01:50070. Figure 2.1 shows a snapshot of the HDFS Web UI for the Hadoop cluster we created previously. To be precise, this page belongs to the NameNode and shows some of the meta data it maintains about the HDFS. You can browse the HDFS, view the NameNode logs and check the status of HDFS nodes. You can also see the live and dead HDFS nodes. In our Hadoop cluster, we have two slave nodes, i.e., hadoopNode02 and hadoopNode03. The web page in Fig. 2.1 shows that we have two live nodes, which means that everything is OK. With this, you can confirm that your cluster has been setup correctly. By clicking on the **Live Nodes** hyperlink, you can see which nodes are alive and are part of the HDFS cluster. Figure 2.2 shows the page displaying details of the live nodes. We can see some meta data about each slave node that is alive. By hovering the mouse over the host names of the nodes, you can see their ip addresses in a tool-tip. Clicking on the hostname of a node will take you to the Web UI of the datanode running on that machine. Figure 2.3 shows a snapshot of the Web UI of the datanode

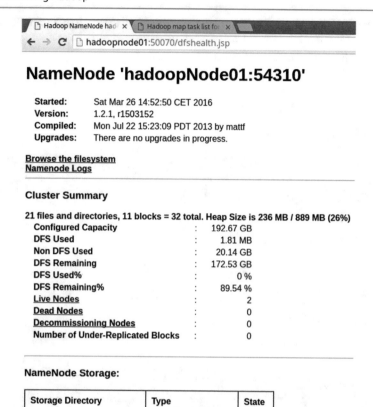

**Fig. 2.1** HDFS Web UI

running on hadoopNode03. The page is divided into two sections. You can use the upper part to browse the HDFS, whereas the lower part can be used to browse Hadoop logs for that machine. It actually allows to browse the logs folder inside the Hadoop installation directory which is the default location for Hadoop logs and can be changed by the user. As we know that on a slave node, in addition to the datanode, the tasktracker also runs, the $HADOOP_HOME/logs directory also stores logs for the tasktracker.

### 2.2.1.2 MapReduce Web UI

The MapReduce Web UI can be found at http://www.<master>:50030 which in our case is http://www.hadoopNode:50030. Browsing to this link will take you to the JobTracker Web UI. To check if all the slaves/TaskTrackers are alive,

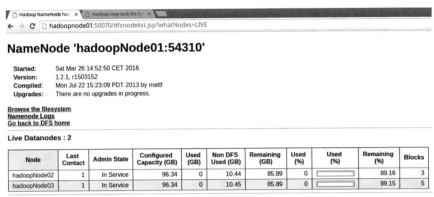

Fig. 2.2 HDFS live nodes page

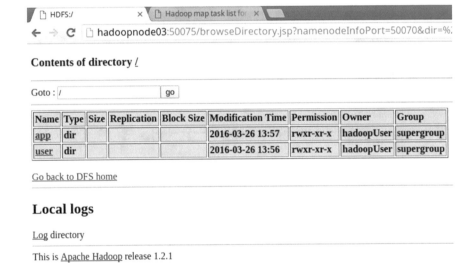

Fig. 2.3 DataNode Web UI

the **Cluster Summary** section should be looked at. Figure 2.4 shows a part of the summary section in which you can see their two nodes alive. If a `TaskTracker` node fails to respond or fails intermittently, it will be put in the columns shown in Fig. 2.5. The **Excluded Nodes** column shows all the nodes that are listed in the `exclude` file in the `conf` directory. Hadoop will not consider these nodes as part of the cluster and will not schedule any tasks to these nodes. In the case of our cluster, we can see that both our slave nodes are alive. Clicking on the Fig. 2.2 in the "Nodes" column will take you to a page which gives more details about the alive nodes or active `TaskTrackers`. You can do the same for Blacklisted, Graylisted,

## Cluster Summary (Heap Size is 298 MB/889 MB)

| Running Map Tasks | Running Reduce Tasks | Total Submissions | Nodes | Occupied Map Slots |
|---|---|---|---|---|
| 0 | 0 | 2 | 2 | 0 |

**Fig. 2.4**  MapReduce cluster summary

| Blacklisted Nodes | Graylisted Nodes | Excluded Nodes |
|---|---|---|
| 0 | 0 | 0 |

**Fig. 2.5**  MapReduce cluster summary-bad nodes

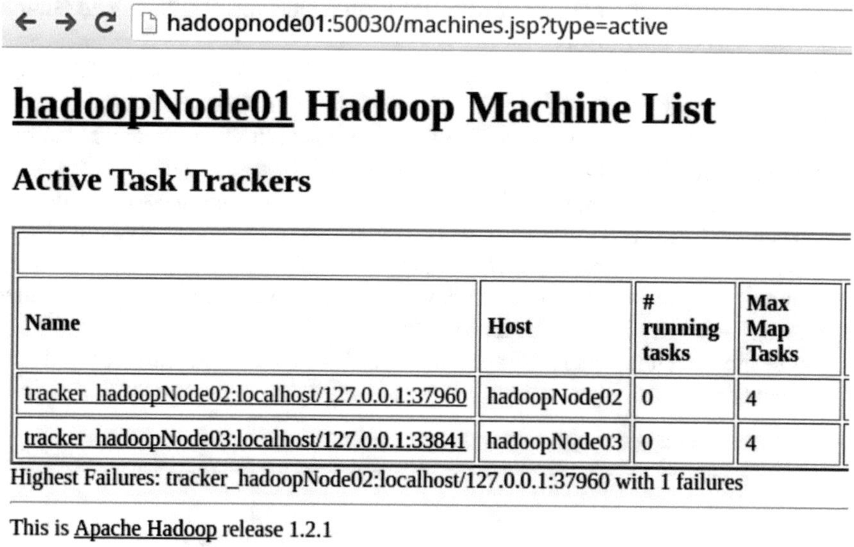

← → C   🗋 hadoopnode01:50030/machines.jsp?type=active

# hadoopNode01 Hadoop Machine List

## Active Task Trackers

| Name | Host | # running tasks | Max Map Tasks |
|---|---|---|---|
| tracker_hadoopNode02:localhost/127.0.0.1:37960 | hadoopNode02 | 0 | 4 |
| tracker_hadoopNode03:localhost/127.0.0.1:33841 | hadoopNode03 | 0 | 4 |

Highest Failures: tracker_hadoopNode02:localhost/127.0.0.1:37960 with 1 failures

This is Apache Hadoop release 1.2.1

**Fig. 2.6**  MR Web UI: active task trackers

and Excluded Nodes. Figure 2.6 shows the active Task Trackers page. We can see that both our slave nodes are running active `TaskTrackers`. Clicking on any name will take you to the Web UI for the corresponding `TaskTracker`. You can also go to the `TaskTracker` page directly by browsing to http://<slave>:50060. On this page you can see the list of running tasks and local logs. Note that the logs' link will point to the logs folder in the Hadoop installation directory on the slave node by default. This means that by clicking on this link, you can also view the logs for the `datanode` running on that node.

At the bottom of the `JobTracker` Web UI, i.e., http://www.hadoopNode01:50030, you can find the **Logs** link. Clicking on it will allow you to browse the logs of the `TaskTracker`. Generally, the node running the `TaskTracker` also runs the `NameNode`, therefore you will also find the `NameNode` logs along with the logs of the `TaskTracker`. You can analyze these logs to find out issues with the Hadoop cluster.

## 2.3  Installing Giraph

1. Go to http://www.giraph.apache.org/releases.html and choose a Giraph release of your choice. At the time of writing, only version `1.1.0` and `1.0.0` are available. As both the releases are stable, we will choose the newer version, i.e., `1.1.0`. Click on the link corresponding to version `1.1.0`. This will take you to a page showing different mirror websites from which you can download Giraph. Choose any of the mirrors.
2. Figure 2.7 shows different download options for Giraph. The last three links allow you to download the source code for Giraph in different compression formats, i.e., `tar.bz2`, `tar.gz`, and `zip`. The first two links allow you to download pre-compiled binaries of Giraph. The file `giraph-dist-1.1.0-bin.tar.gz` contains Giraph pre-compiled for `hadoop-1.2.1`, whereas the second file with the name `giraph-dist-1.1.0-hadoop2-bin.tar.gz` contains Giraph pre-compiled for `hadoop-2.5.1`. As we installed `hadoop-1.2.1` in the previous section, we would like to have a compatible Giraph release. Download `giraph-dist-1.1.0-bin.tar.gz`.
3. Go to the directory where Giraph has been downloaded and extract the downloaded compressed file:

```
tar xzf giraph-dist-1.1.0-bin.tar.gz
```

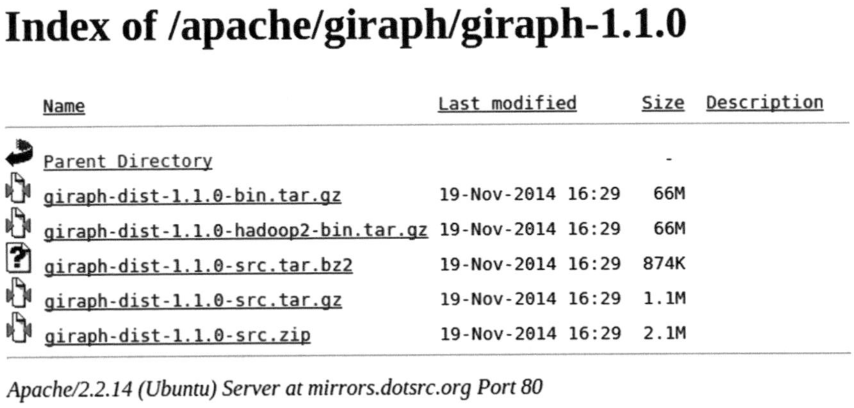

# Index of /apache/giraph/giraph-1.1.0

| Name | Last modified | Size | Description |
|------|---------------|------|-------------|
| Parent Directory | | - | |
| giraph-dist-1.1.0-bin.tar.gz | 19-Nov-2014 16:29 | 66M | |
| giraph-dist-1.1.0-hadoop2-bin.tar.gz | 19-Nov-2014 16:29 | 66M | |
| giraph-dist-1.1.0-src.tar.bz2 | 19-Nov-2014 16:29 | 874K | |
| giraph-dist-1.1.0-src.tar.gz | 19-Nov-2014 16:29 | 1.1M | |
| giraph-dist-1.1.0-src.zip | 19-Nov-2014 16:29 | 2.1M | |

*Apache/2.2.14 (Ubuntu) Server at mirrors.dotsrc.org Port 80*

**Fig. 2.7**  Giraph download options

4. Move the Giraph directory to the `/usr/local` directory:

```
sudo  mv  giraph -1.1.0 - for - hadoop -1.2.1  / usr / local
    /
```

5. Make `hadoopUser` the owner of `giraph` directory and switch to `hadoop User`.

```
sudo  chown  -R  hadoopUser : hadoopGroup  / usr / local /
    giraph -1.1.0 - for - hadoop -1.2.1
su  hadoopUser
```

6. Edit `.bashrc` file and add the following which creates an environment variable `$GIRAPH_HOME` pointing to the Giraph installation directory and adds `giraph` executable script to the `$PATH` variable.

```
export  GIRAPH_HOME =/ usr / local / giraph -1.1.0 - for -
    hadoop -1.2.1
export  PATH =$ PATH : $ GIRAPH_HOME / bin
```

7. Reload the `.bashrc` file.

```
source  ~/. bashrc
```

8. Now you should be able to use the `giraph` script without specifying its absolute path. Try running the script to verify

```
giraph
```

You should see the following output:

```
Usage :  giraph  [-D< Hadoop  property >]  < jar
    containing  vertex >  < parameters  to  jar >
At  a  minimum  one  must  provide  a  path  to  the  jar
    containing  the  vertex  to  be  executed .
```

Keep in mind that the `giraph` script uses the `$HADOOP_HOME` variable to determine which Hadoop installation to use for running the job. If you want to run Giraph on a different Hadoop installation, modify the `$HADOOP_HOME` variable before executing the `giraph` script.

## 2.4   Installing Giraph from Source Code

If you already have a Hadoop cluster running and cannot find a Giraph binary compatible with your Hadoop version, you can compile Giraph from source to get a compatible binary. Follow the steps below to create a Giraph binary compatible with your Hadoop version:

1. Follow steps 2.3 and 1, but instead of downloading a binary download the source distribution which in our example case is `giraph-dist-1.1.0-src.tar.gz`.

2. Go to the directory where the Giraph source has been downloaded and extract the downloaded compressed file:

```
tar xzf giraph-dist-1.1.0-src.tar.gz
```

3. Go to the extracted directory and compile Giraph using the following command:

```
mvn package -Dhadoop.version=1.2.1 -DskipTests
```

`-Dhadoop.version=1.2.1` specifies the version of Hadoop we want to compile Giraph for. The respective jars will be created in the `target` directory of the projects' directories.

4. For later versions of Hadoop which support `yarn` as a resource scheduler, we can add `yarn` support to Giraph using the following command:

```
mvn compile -Phadoop_yarn -Dhadoop.version=2.6.1
    -DskipTests
```

If the above command does not succeed, modify the `pom.xml` file in the source directory and replace the following tag:

```
<munge.symbols>PURE_YARN,STATIC_SASL_SYMBOL</
    munge.symbols>
```

with:

```
<munge.symbols>PURE_YARN</munge.symbols>
```

and recompile Giraph source again.

5. For custom HBase support for a particular version, use the `-Ddep.hbase.version`:

```
mvn compile -Phadoop_yarn -Dhadoop.version=2.6.1
    -Ddep.hbase.version=0.98 -DskipTests
```

6. Many other dependencies' versions can be specified. Go through the `pom.xml` file in the root directory of Giraph source to explore other dependencies. A few of the most important ones are listed below (Table 2.1):

7. Follow steps 3 onwards to install Giraph.

**Table 2.1**  Giraph dependency options

| dep.avro.version |
| --- |
| dep.accumulo.version |
| dep.guava.version |
| dep.hive.version |
| dep.hcatalog.version |
| dep.zookeeper.version |

## 2.5  Running an Example Giraph Job

We can test if Giraph has been deployed successfully along with its dependencies by running an example Giraph job. Running a Giraph job on any Hadoop installation mode is the same. The only difference is that in local mode, we do not have any HDFS setup thus we use the local file system to read the input file and write the output. For clarity, we explain the procedure for local mode separately.

### 2.5.1  Hadoop Local Mode

Follow the following steps to run the shortest path algorithm which comes packaged in `giraph-examples` jar on a local Hadoop installation.

1. Create a file `/tmp/inputGraph.txt` and add the following graph to it. The graph follows the format `[source_id,source_value,[[dest_id, edge_value],...]]`.

```
[0,0,[[1,1],[3,3]]]
[1,0,[[0,1],[2,2],[3,1]]]
[2,0,[[1,2],[4,4]]]
[3,0,[[0,3],[1,1],[4,4]]]
[4,0,[[3,4],[2,4]]]
```

2. Use the `giraph` script to run the shortest path example contained in `giraph-examples-1.1.0.jar`.

```
giraph $GIRAPH_HOME/giraph-examples-1.1.0.jar
    org.apache.giraph.examples.
    SimpleShortestPathsComputation -vif org.
    apache.giraph.io.formats.
    JsonLongDoubleFloatDoubleVertexInputFormat -
    vip /tmp/inputGraph.txt -vof org.apache.
    giraph.io.formats.IdWithValueTextOutputFormat
    -op /tmp/shortestpaths -w 1 -ca giraph.
    SplitMasterWorker=false
```

**Table 2.2**  Sample Giraph job parameters

| Parameter | Description |
|-----------|-------------|
| -vif | Vertex Input Format |
| -vof | Vertex Output Format |
| -vip | Vertex Input Path (Path to input graph) |
| -op | Output path (Path to the directory where output should be stored) |
| -w | Number of workers to use for computation |
| -ca | Custom Arguments |

The first parameter of the `giraph` script should be the jar file containing the vertex. The second parameter should be the vertex class that needs to be executed. The rest of the parameters are described in Table 2.2.

3. As Hadoop runs in a single JVM process in the local mode, always use one worker and force Giraph not to run the Master task in a separate process by setting the custom argument `giraph.SplitMasterWorker` to false. We cannot run the Master separately as we only have one worker available in the local mode.

4. To view the output of the algorithm, use the following command:

```
less /tmp/shortestpaths/part-m-00000
```

You should see the following output:

```
0          1.0
4          5.0
2          2.0
1          0.0
3          1.0
```

The first column of the output represents the vertex IDs whereas the second column represents the distance from vertex with ID 1.

5. If you run the shortest path algorithm again, you will get the following exception:

```
Exception in thread "main" org.apache.hadoop.
    mapred.FileAlreadyExistsException: Output
    directory /user/hadoopUser/output/
    shortestpaths already exists
```

This is because Hadoop does not overwrite a directory that already exists, for its output. You should either specify a different directory using the `-op` parameter or delete the existing directory using the following command:

```
rm -r /tmp/shortestpaths
```

## 2.5.2   Pseudo-Distributed and Clustered Hadoop Mode

Follow the following steps to run the shortest path algorithm which comes packaged in `giraph-examples` jar on a pseudo-distributed or Cluster Hadoop installation. Make sure that the HDFS and MapReduce are up and running. If not, use `start-dfs.sh` script to start the HDFS and `start-mapred.sh` script to start the MapReduce framework.

1. Create a file `/tmp/inputGraph.txt` and add the following graph to it. The graph follows the format `[source_id, source_value, [[dest_id, edge_value],...]]`.

```
[0,0,[[1,1],[3,3]]]
[1,0,[[0,1],[2,2],[3,1]]]
[2,0,[[1,2],[4,4]]]
[3,0,[[0,3],[1,1],[4,4]]]
[4,0,[[3,4],[2,4]]]
```

2. Copy the file to HDFS and verify that the file has been copied. Make sure that HDFS has already been started.

```
hadoop dfs -copyFromLocal /tmp/inputGraph.txt /
    user/hadoopUser/giraphInput/inputGraph.txt
hadoop dfs -ls /user/hadoopUser/giraphInput/
```

3. Use the `giraph` script to run the shortest path example contained in `giraph-examples-1.1.0.jar`.

```
giraph $GIRAPH_HOME/giraph-examples-1.1.0.jar
    org.apache.giraph.examples.
    SimpleShortestPathsComputation -vif org.
    apache.giraph.io.formats.
    JsonLongDoubleFloatDoubleVertexInputFormat -
    vip /user/hadoopUser/giraphInput/inputGraph.
    txt -vof org.apache.giraph.io.formats.
    IdWithValueTextOutputFormat -op /user/
    hadoopUser/output/shortestpaths -w 2
```

The first parameter of the `giraph` script is the jar file containing the vertex. The second parameter is the vertex class that needs to be executed. The rest of the parameters are described in Table 2.2.

4. To view the output of the algorithm, use the following command:

```
hadoop dfs -cat /user/hadoopUser/output/
    shortestpaths/p* | less
```

You should see the following output:

```
0          1.0
4          5.0
2          2.0
1          0.0
3          1.0
```

The first column of the output represents the vertex IDs whereas the second column represents the distance from vertex 0.

5. If you run the shortest path algorithm again, you will get the following exception:

```
Exception in thread "main" org.apache.hadoop.
    mapred.FileAlreadyExistsException: Output
    directory /user/hadoopUser/output/
    shortestpaths already exists
```

This is because Hadoop does not overwrite a directory that already exists, for its output. You should either specify a different directory using the -op parameter or delete the existing directory using the following command:

```
hadoop dfs -rmr /user/hadoopUser/output/
    shortestpaths
```

## 2.6  Monitoring Giraph Application Life Cycle

Giraph offers two interfaces to monitor the life cycle of a Giraph application. You can implement these interfaces once and use the implementations for different jobs. Figure 2.8 shows the life cycle of a Giraph application. The green circles represent different methods that are called between life cycle phases. The boxes represent each phase of the life cycle. The data is loaded into the memory before the application

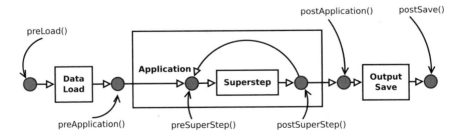

**Fig. 2.8** Giraph application life cycle

starts. Within the application one or more supersteps are executed. Just after the last superstep, the application finishes and the data is saved to the output destination. The interfaces and their scope are explained below.

### 2.6.1  `MasterObserver`

Listing 2.1 shows the source code for the `MasterObserver` interface. The methods of the interface and when they are called are self explanatory. As there is one active master at a time, these methods are executed at one place, i.e., at the master node. The methods of this interface should be used for writing code of global scope, i.e., related to the whole application. The method `applicationFailed()` is called when the application fails. You can override this method and send an email to the application owner with the stack-trace of the application. You can keep trace of the number of supersteps already executed and trigger an action after each superstep. The implementation of these methods totally depend on how the user wants to monitor the Giraph application or trigger some actions. There can be more than one master observer. The `MasterObserver` implementations can be specified by the Giraph property `giraph.master.observers`. It can also be specified using the `addMasterObserverClass()` method of the `GiraphConfiguration` object.

### 2.6.2  `WorkerObserver`

Listing 2.2 shows the source code for the `WorkerObserver` interface. During execution of a Giraph application, there as many instances of `WorkerObservers` as there are workers. Thus methods of this interface should not be used for writing code of global scope. The methods can be used to keep track of individual workers. Implementations of the `WorkerObserver` interface can be specified using the Giraph option `giraph.worker.observers` or by the `addWorkerObserverClass()` method of the `GiraphConfiguration` object.

## 2.7   Monitoring Giraph Jobs

Giraph jobs are executed just like normal Hadoop jobs. This means that you can monitor a Giraph job just like you would monitor a Hadoop job. You can do that either through command line or by using Hadoop's Web UI.

Listing 2.1: MasterObserver.java

```java
public interface MasterObserver extends
    ImmutableClassesGiraphConfigurable {
  /**
   * Before application begins.
   */
  void preApplication();

  /**
   * After application ends.
   */
  void postApplication();

  /**
   * If there is an error during the application.
   *
   * @param e Exception that caused failure. May be
   *     null.
   */
  void applicationFailed(Exception e);

  /**
   * Before each superstep starts.
   *
   * @param superstep The superstep number
   */
  void preSuperstep(long superstep);

  /**
   * After each superstep ends.
   *
   * @param superstep The superstep number
   */
  void postSuperstep(long superstep);
}
```

### 2.7.1  Using Hadoop Commands to Monitor Giraph Jobs

As we know that Giraph jobs are actually Hadoop jobs, when a job has been submitted to the Hadoop cluster, we can monitor the job using Hadoop commands on the command line.

**hadoop job -list**

This command lists the currently running Hadoop jobs. Figure 2.9 shows a sample output of this command.

**hadoop job -status <jobId>**

Listing 2.2: WorkerObserver.java

```java
public interface WorkerObserver extends
    ImmutableClassesGiraphConfigurable {
  /**
   * Initialize the observer. This method is
      executed once on each worker before
   * loading.
   */
  void preLoad();

  /**
   * Initialize the observer. This method is
      executed once on each worker after
   * loading before the first superstep starts.
   */
  void preApplication();

  /**
   * Execute the observer. This method is executed
      once on each worker before
   * each superstep starts.
   *
   * @param superstep number of superstep
   */
  void preSuperstep(long superstep);

  /**
   * Execute the observer. This method is executed
      once on each worker after
   * each superstep ends.
   *
   * @param superstep number of superstep
   */
  void postSuperstep(long superstep);

  /**
   * Finalize the observer. This method is executed
      once on each worker after
   * the last superstep ends before saving.
   */
  void postApplication();

  /**
   * Finalize the observer. This method is executed
      once on each worker after
   * saving.
   */
  void postSave();
}
```

```
1 jobs currently running
JobId    State    StartTime    UserName Priority    SchedulingInfo
job_201604161310_0002    1    1460806314615    hadoopUser  NORMAL NA
```

**Fig. 2.9** Output of `hadoop job -list` command

```
Job: job 201604161310 0002 file: hdfs://hadoopNode01:54310/app/hadoop/tmp/mapred/
staging/hadoopUser/.staging/job_201604161310_0002/job.xml
tracking URL: http://hadoopNode01:50030/jobdetails.jsp?jobid=job 201604161310 0002
map() completion: 1.0
reduce() completion: 1.0
Failure Info: # of failed Map Tasks exceeded allowed limit.
FailedCount: 1.
LastFailedTask: task 201604161310 0002 m 000001
Counters: 6
    Job Counters
        Launched map tasks=3
        SLOTS MILLIS REDUCES=0
        Total time spent by all reduces waiting after reserving slots (ms)=0
        Failed map tasks=1
        SLOTS MILLIS MAPS=56482
        Total time spent by all maps waiting after reserving slots (ms)=0
```

**Fig. 2.10** Output of `hadoop job -status` command

This command shows the status of a job. The following command checks the status of the job listed above as a result of `hadoop job -list` command:

```
hadoop job -status job_201604161310_0002
```

Figure 2.10 shows the output of the above command. Note that the job failed. To inquire further the reason for the failure, we can use the `hadoop job history` command.

### hadoop job -history <job-output-dir>

This command is used to view execution details of a job when it has finished successfully or failed. We have to specify the output directory of a job to view its history. Figure 2.11 shows the output of this command for our example job. It shows the number of launched, successful, failed, and killed tasks. It also shows the stack-trace of the error causing the job failure.

### hadoop job -kill <job-id>

This command is used to kill a running job. It might be useful to kill a job when it is taking too much more time than expected or if you got some parameters of the job wrong.

### hadoop job -kill-task <task-id>

This command is used to kill a running task. If a task is taking more time than normal and we suspect that there is a problem, we can kill the task ourselves rather than waiting for Hadoop to kill it. Killing a task is not counted against failed attempts.

```
Hadoop job: 0002_1460806314615_hadoopUser
==========================================
Job tracker host name: job
Job tracker start time: Sat May 22 10:09:21 CET 2016
User: hadoopUser
JobName: Giraph: org.apache.giraph.examples.SimpleShortestPathsComputation
JobConf: hdfs://hadoopNode01:54319/app/hadoop/tmp/mapred/staging/hadoopUser/.staging/job_2016
04161310_0002/job.xml
Submitted At: 16-Apr-2016 13:31:54
Launched At: 16-Apr-2016 13:31:54 (0sec)
Finished At: 16-Apr-2016 13:32:36 (41sec)
Status: FAILED
Counters:

|Group Name    |Counter name   |Map Value  |Reduce Value   |Total Value|
-----------------------------------------------------------------------
============================

Task Summary
====================
Kind     Total   Successful  Failed  Killed  StartTime       FinishTime

Setup    1       1           0       0       16-Apr-2016 13:32:09 16-Apr-2016 13:32:11 (2sec)
Map      3       0           1       2       16-Apr-2016 13:32:11 16-Apr-2016 13:32:35 (23sec)
Reduce   0       0           0       0
Cleanup  1       1           0       0       16-Apr-2016 13:32:33 16-Apr-2016 13:32:35 (1sec)
====================

No Analysis available as job did not finish

FAILED SETUP task list for 0002_1460806314615_hadoopUser
TaskId  StartTime   FinishTime  Error
==================================

KILLED SETUP task list for 0002_1460806314615_hadoopUser
TaskId  StartTime   FinishTime  Error
==================================

FAILED MAP task list for 0002_1460806314615_hadoopUser
TaskId  StartTime   FinishTime  Error   InputSplits
==================================
task_201604161310_0002_m_000001 16-Apr-2016 13:32:12 16-Apr-2016 13:32:33 (21sec)
java.lang.IllegalStateException
    at org.apache.giraph.graph.GraphMapper.run(GraphMapper.java:194)
    at org.apache.hadoop.mapred.MapTask.runNewMapper(MapTask.java:764)
    at org.apache.hadoop.mapred.MapTask.run(MapTask.java:364)
    at org.apache.hadoop.mapred.Child$4.run(Child.java:255)
    at java.security.AccessController.doPrivileged(Native Method)
    at javax.security.auth.Subject.doAs(Subject.java:422)
    at org.apache.hadoop.security.UserGroupInformation.doAs(UserGroupInformation.java:1190)
    at org.apache.hadoop.mapred.Child.main(Child.java:249)
Caused by: java.lang.RuntimeException: java.net.UnknownHostException: hadoopnode03: unknown error
    at org.apache.giraph.bsp.BspService.&lt;init&gt;(BspService.java:358)
    at org.apache.giraph.worker.BspServiceWorker.&lt;init&gt;(BspServiceWorker.java:195)
    at org.apache.giraph.graph.GraphTaskManager.instantiateBspService(GraphTaskManager.java:607)
    at org.apache.giraph.graph.GraphTaskManager.setup(GraphTaskManager.java:239)
    at org.apache.giraph.graph.GraphMapper.setup(GraphMapper.java:61)
    at org.apache.giraph.graph.GraphMapper.run(GraphMapper.java:91)
    ... 7 more
Caused by: java.net.UnknownHostException: hadoopnode03: unknown error
```

**Fig. 2.11** Output of `hadoop job -history <job-output-dir>` command

## `hadoop job -set-priority <job-id> <priority>`

If multiple jobs are running on the cluster, we can assign priorities to a job by this command. Allowed priority values are VERY_HIGH, HIGH, NORMAL, LOW, VERY_LOW.

## 2.7.2  Using Hadoop UI to Monitor Giraph Jobs

We can monitor Giraph jobs by browsing to the JobTracker UI. If you have set up the Hadoop cluster as described in this book, you can browse to http://www. hadoopnode01:50030/jobtracker.jsp for accessing the JobTracker UI. Click on the ID of the job of which you want to view the status. The job ids are listed in different sections based on their status if they are running, failed, or finished successfully. When you click on the job id, it takes you to a page which gives the details about the number of Map and Reduce tasks running, completed, failed, or killed (Fig. 2.12. The page also shows you the counters of the job and path to the configuration file which was used to execute the job. If you want to see the details about all the tasks, click on "map" in the first column of the task table which will show the status of each task (Fig. 2.13). Similarly, if you want to find out details about the killed tasks, click on the number in the "killed" column. On the next page, you will see all tasks that were killed. Clicking on the id of a task will take you to the page where you can see all the task attempts (Fig. 2.14). Here you can access the logs of the killed task to find out the problem.

## 2.8  Configuring Giraph

Giraph is very much tunable just like many other big data processing systems. Its behavior can be modified to fit the algorithmic needs or to improve performance. We explained some of the Giraph parameters in Sect. 2.5. Many more parameters/configuration options will be explained throughout this book. You can visit Sect. 2.8.4 to get the list of configuration described in this book.

In this section, we will explain the different ways in which Giraph can be configured. The choice of method depends on the use case. Figure 2.15 shows the three different methods in which Giraph can be configured and their hierarchy.

## 2.8.1  Giraph-Specific Options

The configuration file `giraph-site.xml` is located in `$GIRAPH_HOME/conf` directory. It contains the **Giraph-specific** global configuration that should be set across runs. Any configuration options set in this file can still be overridden by job specific or algorithm-specific options. The layout of this file is similar to Hadoop configuration files, e.g., `core-site.xml`. The configuration that will not change often between different job runs, should be put in this file.

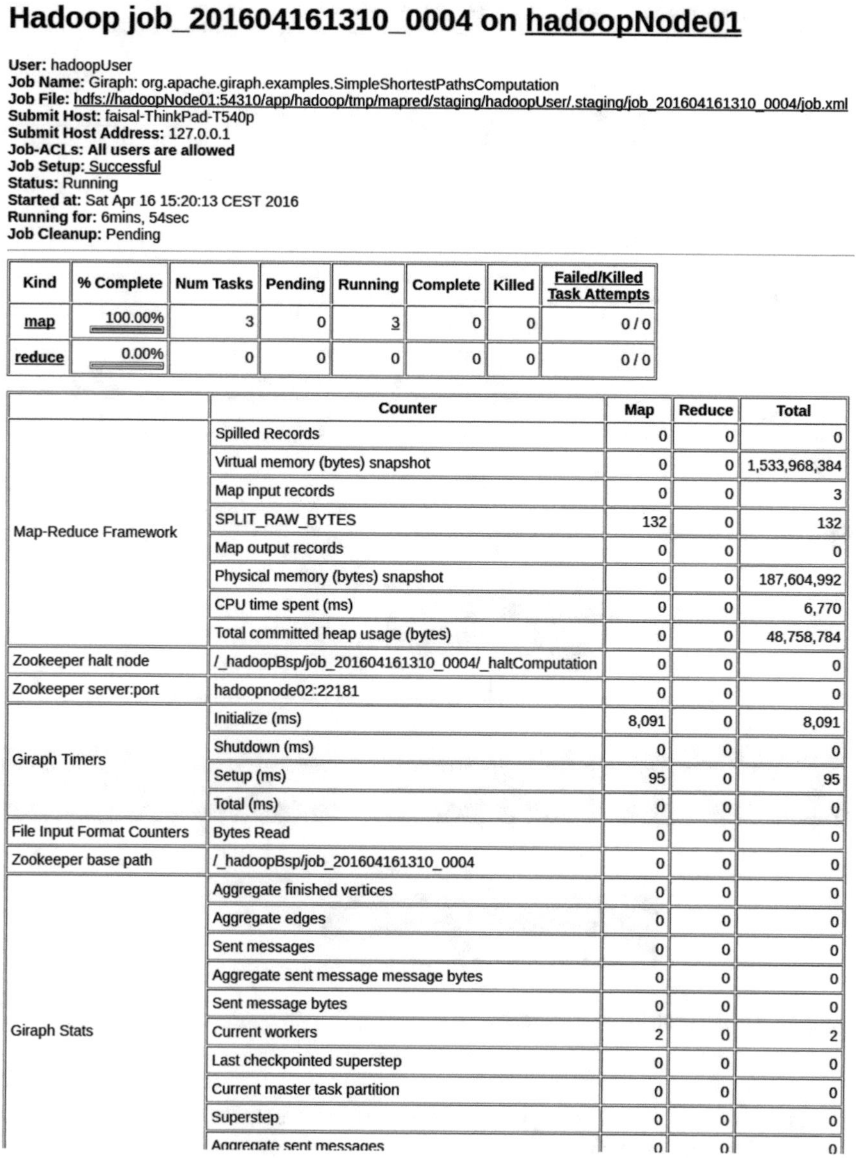

# Hadoop job_201604161310_0004 on <u>hadoopNode01</u>

**User:** hadoopUser
**Job Name:** Giraph: org.apache.giraph.examples.SimpleShortestPathsComputation
**Job File:** <u>hdfs://hadoopNode01:54310/app/hadoop/tmp/mapred/staging/hadoopUser/.staging/job_201604161310_0004/job.xml</u>
**Submit Host:** faisal-ThinkPad-T540p
**Submit Host Address:** 127.0.0.1
**Job-ACLs: All users are allowed**
**Job Setup:** <u>Successful</u>
**Status:** Running
**Started at:** Sat Apr 16 15:20:13 CEST 2016
**Running for:** 6mins, 54sec
**Job Cleanup:** Pending

| Kind | % Complete | Num Tasks | Pending | Running | Complete | Killed | Failed/Killed Task Attempts |
|------|-----------|-----------|---------|---------|----------|--------|------------------------------|
| <u>map</u> | 100.00% | 3 | 0 | 3 | 0 | 0 | 0 / 0 |
| <u>reduce</u> | 0.00% | 0 | 0 | 0 | 0 | 0 | 0 / 0 |

| | Counter | Map | Reduce | Total |
|--|---------|-----|--------|-------|
| | Spilled Records | 0 | 0 | 0 |
| | Virtual memory (bytes) snapshot | 0 | 0 | 1,533,968,384 |
| | Map input records | 0 | 0 | 3 |
| Map-Reduce Framework | SPLIT_RAW_BYTES | 132 | 0 | 132 |
| | Map output records | 0 | 0 | 0 |
| | Physical memory (bytes) snapshot | 0 | 0 | 187,604,992 |
| | CPU time spent (ms) | 0 | 0 | 6,770 |
| | Total committed heap usage (bytes) | 0 | 0 | 48,758,784 |
| Zookeeper halt node | /_hadoopBsp/job_201604161310_0004/_haltComputation | 0 | 0 | 0 |
| Zookeeper server:port | hadoopnode02:22181 | 0 | 0 | 0 |
| | Initialize (ms) | 8,091 | 0 | 8,091 |
| Giraph Timers | Shutdown (ms) | 0 | 0 | 0 |
| | Setup (ms) | 95 | 0 | 95 |
| | Total (ms) | 0 | 0 | 0 |
| File Input Format Counters | Bytes Read | 0 | 0 | 0 |
| Zookeeper base path | /_hadoopBsp/job_201604161310_0004 | 0 | 0 | 0 |
| | Aggregate finished vertices | 0 | 0 | 0 |
| | Aggregate edges | 0 | 0 | 0 |
| | Sent messages | 0 | 0 | 0 |
| | Aggregate sent message message bytes | 0 | 0 | 0 |
| | Sent message bytes | 0 | 0 | 0 |
| Giraph Stats | Current workers | 2 | 0 | 2 |
| | Last checkpointed superstep | 0 | 0 | 0 |
| | Current master task partition | 0 | 0 | 0 |
| | Superstep | 0 | 0 | 0 |
| | Aggregate sent messages | 0 | 0 | 0 |

**Fig. 2.12**  Giraph job details

## 2.8.2   Job-Specific Options

Job-specific options are the ones that are provided at the command line to the driver
class while running a Giraph job. These parameters override the options specified in

**Hadoop map task list for** <u>job_201604161310_0004</u> **on** <u>hadoopNode01</u>

**Running Tasks**

| Task | Complete | Status | Start Time | Finish Time | Errors | Counters |
|------|----------|--------|------------|-------------|--------|----------|
| task_201604161310_0004_m_000000 | 100.00% | MASTER_ZOOKEEPER_ONLY - 0 finished out of 2 on superstep -1 | 16-Apr-2016 15:20:29 | | | 30 |
| task_201604161310_0004_m_000001 | 100.00% | startSuperstep: WORKER_ONLY - Attempt=0, Superstep=-1 | 16-Apr-2016 15:20:29 | | | 15 |
| task_201604161310_0004_m_000002 | 100.00% | startSuperstep: WORKER_ONLY - Attempt=0, Superstep=-1 | 16-Apr-2016 15:20:29 | | | 15 |

Go back to JobTracker

This is Apache Hadoop release 1.2.1

**Fig. 2.13**   Running map tasks

**Job** <u>job_201604161310_0004</u>

**All Task Attempts**

| Task Attempts | Machine | Status | Progress | Start Time | Finish Time | Errors | Task Logs | Counters | Actions |
|---------------|---------|--------|----------|------------|-------------|--------|-----------|----------|---------|
| attempt_201604161310_0004_m_000000_0 | /default-rack/hadoopNode02 | RUNNING | 100.00% | 16-Apr-2016 15:20:29 | | | Last 4KB Last 8KB All | 0 | |

**Input Split Locations**

Go back to the job
Go back to JobTracker

This is Apache Hadoop release 1.2.1

**Fig. 2.14**   Task attempts

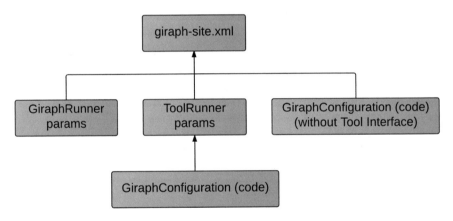

**Fig. 2.15**   Giraph configuration hierarchy

`giraph-site.xml`. These options can be set in different ways depending on the type of driver you are using.

### 2.8.2.1   Using `GiraphRunner` as Driver

`GiraphRunner` can be used as a driver for construction of Giraph jobs. `Giraph Runner` is a helper class to run Giraph applications by specifying the actual class name to use (i.e., vertex, vertex input/output format, combiner, etc.). `Giraph Runner` is the default entry point for Giraph jobs running on any Hadoop cluster, MRv1 or v2, including Hadoop-specific configuration and setup. To find out a list of configuration options supported by the `giraph` script, run the following command using `hadoopUser`:

```
giraph org.apache.giraph.GiraphRunner -h
```

The output of the above command will not list all the options supported by Giraph but only those options which are most relevant to a specific job run. The options are also abbreviated for making them easier to specify on the command line. For instance, to specify the vertex input format, instead of specifying the complete property name `giraph.vertexInputFormatClass`, we can just use the parameter specifier `-vif`. For configuration options that are not listed specifically in the output of the above command, `-ca` or `-customArguments` parameter specifier can be used. For instance, in step 2 in Sect. 2.5.1, the option `giraph.SplitMasterWorker=false` is specified after `-ca` and is a custom argument.

### 2.8.2.2 Using a Custom Driver Implementing `Tool` Interface

If you have written your own driver using the `Tool` interface, command line Giraph options can be specified using the `-D` parameter specifier. Note that you cannot use the abbreviated property specifiers as they are specific to `GiraphRunner`. You will have to specify the complete property names for setting them. If your driver class is `myorg.driverTool`, you can specify job-specific Giraph parameters and run the job as follows:

```
giraph myorg.driverTool \
-D giraph.vertexInputFormatClass=org.apache.
    giraph.io.formats.
    JsonLongDoubleFloatDoubleVertexInputFormat \
-D giraph.vertex.input.dir="/home/User/git/
    GiraphDemoRunnerMaven/src/main/resources/
    input.txt" \
-D giraph.minWorkers=1 \
-D giraph.maxWorkers=1 \
-D giraph.SplitMasterWorker=false \
-D giraph.localTestMode=true \
-D mapred.output.dir="/home/User/git/
    GiraphDemoRunnerMaven/src/main/resources/
    output" \
-D giraph.vertexOutputFormatClass="org.apache.
    giraph.io.formats.
    IdWithValueTextOutputFormat"
```

### 2.8.3 Algorithm-Specific Options

If there are some Giraph options that are required to be set specific to the type of the algorithm, they can be set from within the code using the `GiraphConfiguration` object. The options set using this object override the options in `giraph-site.xml` and the ones provided as parameters to `GiraphRunner`. Note that you will have to write your own driver program to fix some Giraph options from within the code.

### 2.8.4  Giraph Configuration Summary

We summarize in Table 2.3 the list of Giraph configurations described and their location in this book. We only consider the most important configurations, however you should visit http://www.giraph.apache.org/options.html for the complete list of configuration options.

## 2.9  Creating a Giraph IDE Project

In this section, we will be explaining how to create a Giraph project in Eclipse and IntelliJ IDEA, the two most famous integrated development environments (IDEs) for Java development. For each IDE, we will explain project creation with and without Maven. You can skip this section and download the readymade project files from the Github repository of this book.[1] We will run the simple shortest path algorithm that comes packaged in **giraph-examples** jar. This algorithm takes as input a graph and finds shortest path distance from vertex with ID 1 to all other vertexes. Create the file `/tmp/tiny_graph.txt` and add the graph in Listing 2.3.

Listing 2.3: "Tiny Graph"

```
[0,0,[[1,1],[3,3]]]
[1,0,[[0,1],[2,2],[3,1]]]
[2,0,[[1,2],[4,4]]]
[3,0,[[0,3],[1,1],[4,4]]]
[4,0,[[3,4],[2,4]]]
```

The graph is in JSON format where each line represents a vertex in the following format:

```
[source_id,source_value,[[dest_id, edge_value
    ],...]]
```

This file will be used as an input for the shortest path algorithm for which we create the a project using different IDEs, in the following sections.

---

[1] The source code implementation of the projects is available on https://www.github.com/sakrsherif/GiraphBookSourceCodes/tree/master/chapter02.

**Table 2.3**  A summary of Giraph's configurations described in this book

| Name | Description | Section |
|---|---|---|
| giraph.SplitMasterWorker | Enable/disable running the master in a separate worker | 2.8.2.1 |
| giraph.vertexInputFormatClass | Specify the vertex input class | 3.5.4 |
| giraph.edgeInputFormatClass | Specify the edge input class | 3.5.5 |
| giraph.vertexOutputFormatClass | Specify the vertex output class | 3.6.1 |
| giraph.edgexOutputFormatClass | Specify the edge output class | 3.6.2 |
| giraph.vertexOutputFormatThreadSafe | Assert that the vertex output format is thread safe | 5.3.1 |
| giraph.useInputSplitLocality | Read local graph splits first | 5.3.1 |
| giraph.numComputeThreads | Number of computation threads at each worker | 5.3.1 |
| giraph.numInputThreads | Number of threads for loading a graph split | 5.3.1 |
| giraph.numOutputThreads | Number of threads for output writing | 5.3.1 |
| giraph.messageEncodeAndStoreType | Message exchange optimization | 5.3.2 |
| giraph.useBigDataIOForMessages | Enable/disable large message storage for more then 4 Billion messages | 5.3.2 |
| giraph.masterComputeClass | Implementation of MasterCompute | 5.1.1 |
| giraph.aggregatorWriterClass | Implementation of the aggregator writer | 5.1.2.3 |
| giraph.textAggregatorWriter.filename | Aggregator writer output file | 5.1.2.3 |
| giraph.textAggregatorWriter.frequency | Specify when to write the aggregator content to disk | 5.1.2.3 |
| giraph.messageCombinerClass | Implementation of message combiners | 5.1.3 |
| giraph.maxMutationsPerRequest | Control the frequency of synchronizing mutation requests between different workers | 5.2.1 |
| giraph.vertex.resolver.create.on.msgs | Enable/disable creating new vertices when a nonexisting vertex receive a message | 5.2.1 |
| giraph.inputOutEdgesClass | Specify the edge storage class at each vertex during graph loading | 5.3.3 |
| giraph.outEdgesClass | Specify the edge storage class at each vertex | 5.3.3 |
| giraph.graphPartitionerFactoryClass | Define the graph partitioning class | 5.4 |
| giraph.partitionClass | Partition storage class | 5.4 |
| giraph.useOutOfCoreGraph | Store graph partitions on disk | 5.3.4 |
| giraph.maxPartitionsInMemory | Maximum graph partitions in a worker's memory | 5.3.4 |

(continued)

**Table 2.3** (continued)

| Name | Description | Section |
|------|-------------|---------|
| giraph.useOutOfCoreMessages | Store incoming messages on disk | 5.3.4 |
| giraph.maxMessagesInMemory | The maximum number of messages stored in a worker's memory | 5.3.4 |
| giraph.partitionsDirectory | Parition storage path in local file system | 5.3.4 |
| giraph.messagesDirectory | Message storage path in local file system | 5.3.4 |
| giraph.messagesBufferSize | Out-of-core cache size for messages in memory | 5.3.4 |
| giraph.isStaticGraph | Indicate that no edge mutations exists in runtime | 5.3.4 |
| giraph.useSuperstepCounters | Enable/disable superstep counters | 5.6.1 |
| mapreduce.job.counters.limit | Control the maximum number of counters per Hadoop job | 5.6.1 |
| mapred.child.java.opts | Hadoop options for Giraph workers | 5.6.4 |
| giraph.checkpointFrequency | The number of supersteps to wait before writing checkpointing | 5.7.1 |
| giraph.checkpoint.io.threads | Number of threads at a worker to write a checkpoint data to HDFS | 5.7.1 |
| giraph.cleanupCheckpointsAfterSuccess | Delete checkpointing files after runtime | 5.7.1 |
| giraph.checkpoint.compression.codec | Compression algorithm for checkpointing files | 5.7.1 |
| giraph.checkpointDirectory | Target for checkpointing in HDFS | 5.7.1 |
| dfs.replication | The replication factor in HDFS | 5.7.1 |

### 2.9.1 Eclipse

Download the latest **Eclipse** IDE for Java Developers from https://www.eclipse.org/downloads compatible with your operating system. We will be using Linux as the operating system and Eclipse Mars for demonstration which is the latest Eclipse version at the time of writing this book. Make sure to have JDK 1.7 or above on your computer.

1. In Eclipse, go to **File | New | Java Project** to create a basic Java project. As shown in Fig. 2.16, specify a **Project name** (GiraphDemoRunner in our case), select the **JRE** and click **Next**.
2. Accept the default **Java Settings** and click on **Finish** (Fig. 2.17).
3. Download latest stable release (1.1.0 at the time of writing) of **Apache Giraph** from http://www.giraph.apache.org/releases.html. Two different distributions of Giraph can be downloaded: **giraph-dist-1.1.0-bin.tar.gz** which is built with **hadoop-1.2.1** and **giraph-dist-1.1.0-hadoop2-bin.tar.gz** which is built with

**hadoop-2.5.1**. Whatever distribution you choose, make sure to also download the corresponding Hadoop version from https://www.archive.apache.org/dist/ hadoop/core/. We will be using Giraph with hadoop-1.2.1 in this section. Uncompress the downloaded Giraph and Hadoop distributions.

4. Right click on the project **GiraphDemoRunner** in **Package** view in Eclipse and choose **Build Path | Configure Build Path** (Fig. 2.18).

5. In the **Libraries** tab, click on **Add External Jars** and select all the jar files in the uncompressed directory **hadoop-1.2.1/lib**. Click **Add External Jars** again and select **hadoop-core-1.2.1.jar**, **hadoop-client-1.2.1.jar** and **hadoop-tools-1.2.1.jar** from **hadoop-1.2.1** directory (Fig. 2.19).

6. Using the same process above, add Giraph core jar **giraph-core-1.1.0** and examples **giraph-examples-1.1.0.jar** from the Giraph main directory and all the jar files inside the **lib** directory. Click on **OK** button. Now the Giraph project has been set up and we can start writing Giraph applications.

7. Click on the **src** folder in the project and click on **New | Class** (Fig. 2.20).

8. In the **New Java Class** window, enter the name of the new class i.e. **GiraphDemoRunner** and click on **Finish** (Fig. 2.21).

9. A new Java class file **GiraphDemoRunner.java** will be created under the **src** folder in the **Package View**. Also the file will automatically be opened in the editor pane. Listing 2.4 contains the code for simple shortest path algorithm. Copy this code and paste in the Java file.

10. Right-click on the **GiraphDemoRunner.java** in the **Package View** and choose **Run As | Java Application**. If the execution throws an error, follow Sects. 2.10 and 5.6.3. The code after successful execution will produce two output files in the **/tmp/graph_out** directory. The empty file **_SUCCESS** represents the successful execution of the program. The absence of this file means that something went wrong during program execution. The other file **part-m-00000** contains the output of the program.

Listing 2.4: GiraphDemoRunner.java

```
1  import org.apache.giraph.conf.GiraphConfiguration;
2  import org.apache.giraph.examples.
      SimpleShortestPathsComputation;
3  import org.apache.giraph.io.formats.*;
4  import org.apache.giraph.job.GiraphJob;
5  import org.apache.hadoop.conf.Configuration;
6  import org.apache.hadoop.fs.Path;
7  import org.apache.hadoop.mapreduce.lib.output.
      FileOutputFormat;
8  import org.apache.hadoop.util.Tool;
9  import org.apache.hadoop.util.ToolRunner;
```

```java
public class GiraphDemoRunner implements Tool{

    private Configuration conf;
    public Configuration getConf() {
        return conf;
    }
    public void setConf(Configuration conf) {
        this.conf = conf;
    }
    public int run(String[] arg0) throws Exception {
        String inputPath="/tmp/tiny_graph.txt";
        String outputPath="/tmp/graph_out";
        GiraphConfiguration giraphConf = new
            GiraphConfiguration(getConf());
        giraphConf.setComputationClass(
            SimpleShortestPathsComputation.class);
        giraphConf.setVertexInputFormatClass(
            JsonLongDoubleFloatDoubleVertexInputFormat.
            class);
        GiraphFileInputFormat.addVertexInputPath(
            giraphConf, new Path(inputPath));
        giraphConf.setVertexOutputFormatClass(
            IdWithValueTextOutputFormat.class);
        giraphConf.setLocalTestMode(true);
        giraphConf.setWorkerConfiguration(1, 1, 100);

        giraphConf.SPLIT_MASTER_WORKER.set(giraphConf,
            false);
        InMemoryVertexOutputFormat.
            initializeOutputGraph(giraphConf);
        GiraphJob giraphJob = new GiraphJob(giraphConf
            ,"GiraphDemo");
        FileOutputFormat.setOutputPath(giraphJob.
            getInternalJob(), new Path(outputPath));
        giraphJob.run(true);
        return 0;
    }

    public static void main(String[] args) throws
        Exception{
        ToolRunner.run(new GiraphDemoRunner(), args);
    }
}
```

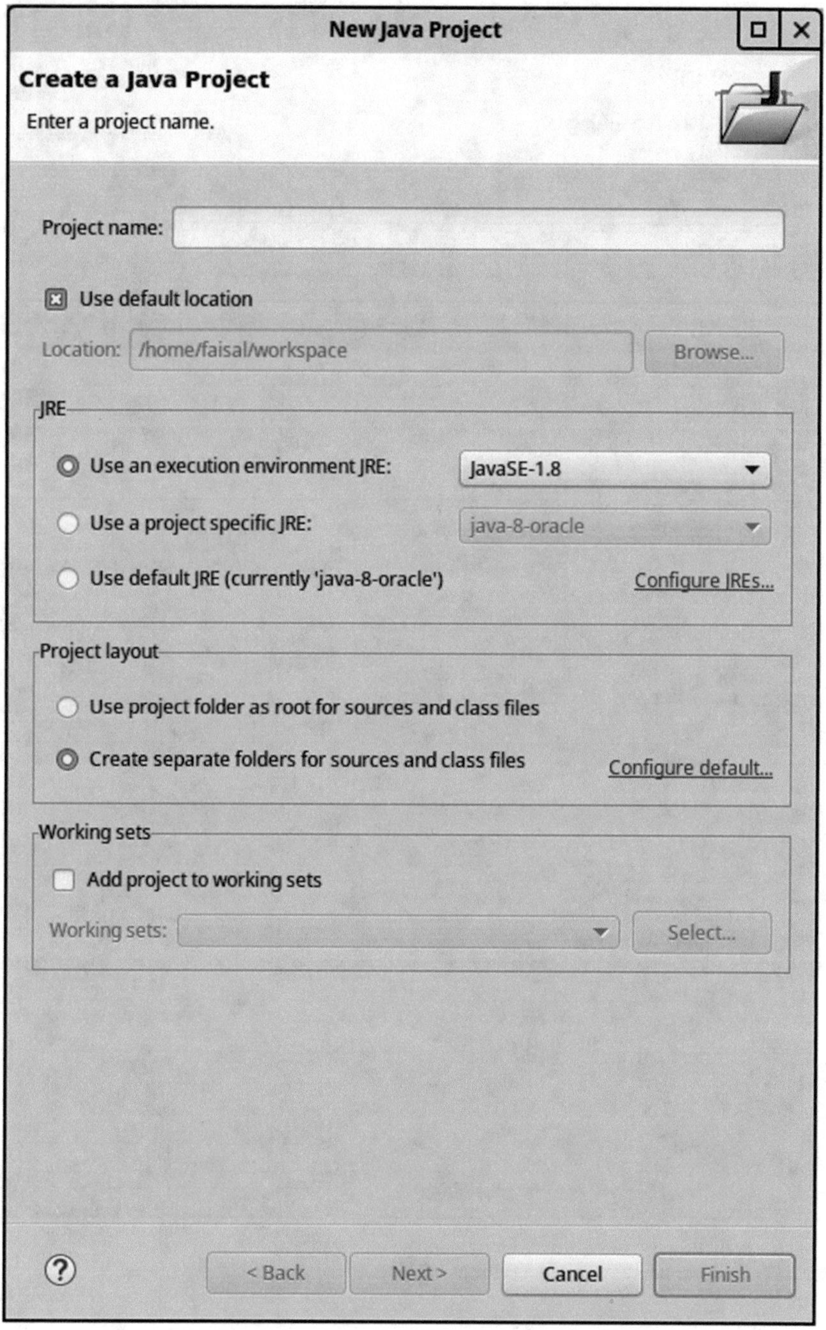

**Fig. 2.16** Eclipse: new Java project window

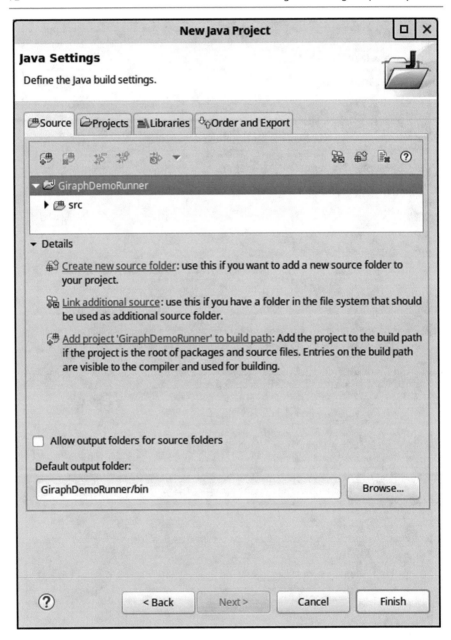

**Fig. 2.17** Eclipse: new Java project settings window

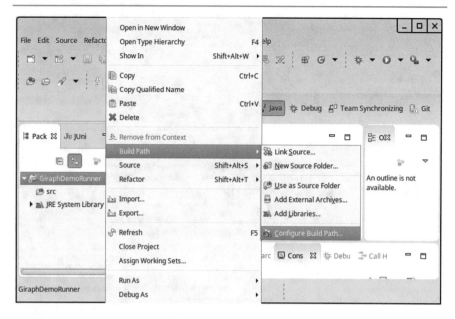

**Fig. 2.18**   Eclipse: build path context menu

## 2.9.2   Eclipse with Maven

Using Maven, we do not need to download and add jar files manually. We can use Maven to specify core project dependencies and it will manage the download of the transitive dependencies automatically. It makes the project portable, light weight and less complex. To demonstrate the process, we will be using Giraph with Hadoop2 and Hadoop version 2.5.1.

1. Download Eclipse and create a Java project as explained in the Sect. 2.9.1. Right click on the **GiraphDemoRunner** project in the package view window on the left and click on **Configure | Convert to Maven Project** (Fig. 2.22).
2. On the **Create new POM** window, accept the defaults and click on **Finish**. This will generate a **pom.xml** file in the main project folder. Opening this file will open a graphical editor. At the bottom of this editor, click on **Dependencies** tab and in the **Dependencies** editor, click on the **Add** button. In the **Select Dependency** window, enter the **Group Id**, **Artifact Id** and **Version** of the **giraph-core-1.1.0-hadoop2.jar** as shown in the Fig. 2.23 and click on the **OK** button.
3. Repeat the process for **giraph-examples-1.1.0-hadoop2.jar**, **hadoop-common-2.5.1.jar** and **hadoop-client-2.5.1.jar**. Alternatively, you can click on the **pom.xml** tab at the bottom of graphical editor and paste the **<dependencies>** XML snippet in the pom.xml file using the sample pom.xml file given in Listing 2.5.
4. Follow the steps from 7 to 10 in Sect. 2.9.1.

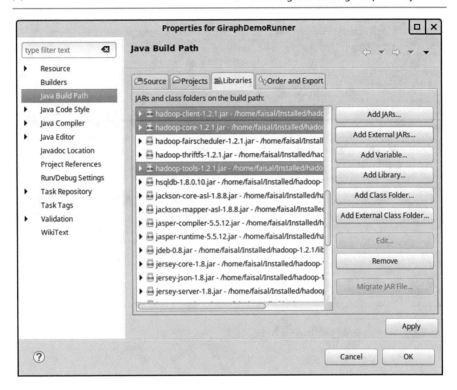

**Fig. 2.19** Eclipse: adding external Jars

5. If you cannot see the output on the console, add a file naming `log4j.properties` in the `src` directory with the following content and run the code again:

```
# Root logger option
log4j.rootLogger=INFO, stdout

# Direct log messages to stdout
log4j.appender.stdout=org.apache.log4j.
    ConsoleAppender
log4j.appender.stdout.Target=System.out
log4j.appender.stdout.layout=org.apache.log4j.
    PatternLayout
log4j.appender.stdout.layout.ConversionPattern=
```

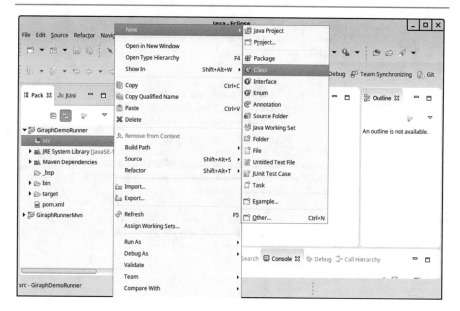

**Fig. 2.20** Eclipse: creating new Java class

---

**Listing 2.5: GiraphDemoRunner pom.xml file**

```xml
<project xmlns="http://maven.apache.org/POM/4.0.0"
    xmlns:xsi="http://www.w3.org/2001/XMLSchema-
    instance" xsi:schemaLocation="http://maven.apache
    .org/POM/4.0.0 http://maven.apache.org/xsd/maven
    -4.0.0.xsd">
  <modelVersion>4.0.0</modelVersion>
  <groupId>GiraphDemoRunner</groupId>
  <artifactId>GiraphDemoRunner</artifactId>
  <version>0.0.1-SNAPSHOT</version>
  <build>
    <sourceDirectory>src</sourceDirectory>
    <plugins>
      <plugin>
        <artifactId>maven-compiler-plugin</
            artifactId>
        <version>3.3</version>
        <configuration>
          <source>1.7</source>
          <target>1.7</target>
        </configuration>
      </plugin>
    </plugins>
  </build>
  <dependencies>
    <dependency>
```

```
21        <groupId>org.apache.giraph</groupId>
22        <artifactId>giraph-core</artifactId>
23        <version>1.1.0-hadoop2</version>
24     </dependency>
25     <dependency>
26        <groupId>org.apache.giraph</groupId>
27        <artifactId>giraph-examples</artifactId>
28        <version>1.1.0-hadoop2</version>
29     </dependency>
30     <dependency>
31        <groupId>org.apache.hadoop</groupId>
32        <artifactId>hadoop-common</artifactId>
33        <version>2.5.1</version>
34     </dependency>
35     <dependency>
36        <groupId>org.apache.hadoop</groupId>
37        <artifactId>hadoop-client</artifactId>
38        <version>2.5.1</version>
39     </dependency>
40
41   </dependencies>
42 </project>
```

### 2.9.3  IntelliJ IDE

Download the latest version of **IntelliJ IDEA** from https://www.jetbrains.com/idea/download/ and install it. Make sure you have already installed **jdk 1.7** or above.

1. Open IntelliJ and go to **File | New | Project...**. This will open a **New Project** window. From the left pane, Select **Java** as the type of the project and select **JDK 1.7** or above from **Project SDK** drop-down menu at the top. If the required JDK is not listed, you can click on the **New...** button next to the drop-down box and select **JDK**. This will open a new window to select the directory of the required JDK. Click on **Next** (Fig. 2.24).
2. In the next window, leave the defaults and click **Next** (Fig. 2.25).
3. In the next window, enter the project name and click on **Finish** (Fig. 2.26).
4. An empty Java project will be created which we can modify to run Giraph programs. Figure 2.27 shows different parts of the IDE.
5. Now, we can add Giraph project dependencies required to run a Giraph program. Click on **File | Project Structure...**. In the **Project Structure** window, click on **Modules** on the left-most pane and then open the **Dependencies** tab. Click on the plus button to add jar files (Fig. 2.28).

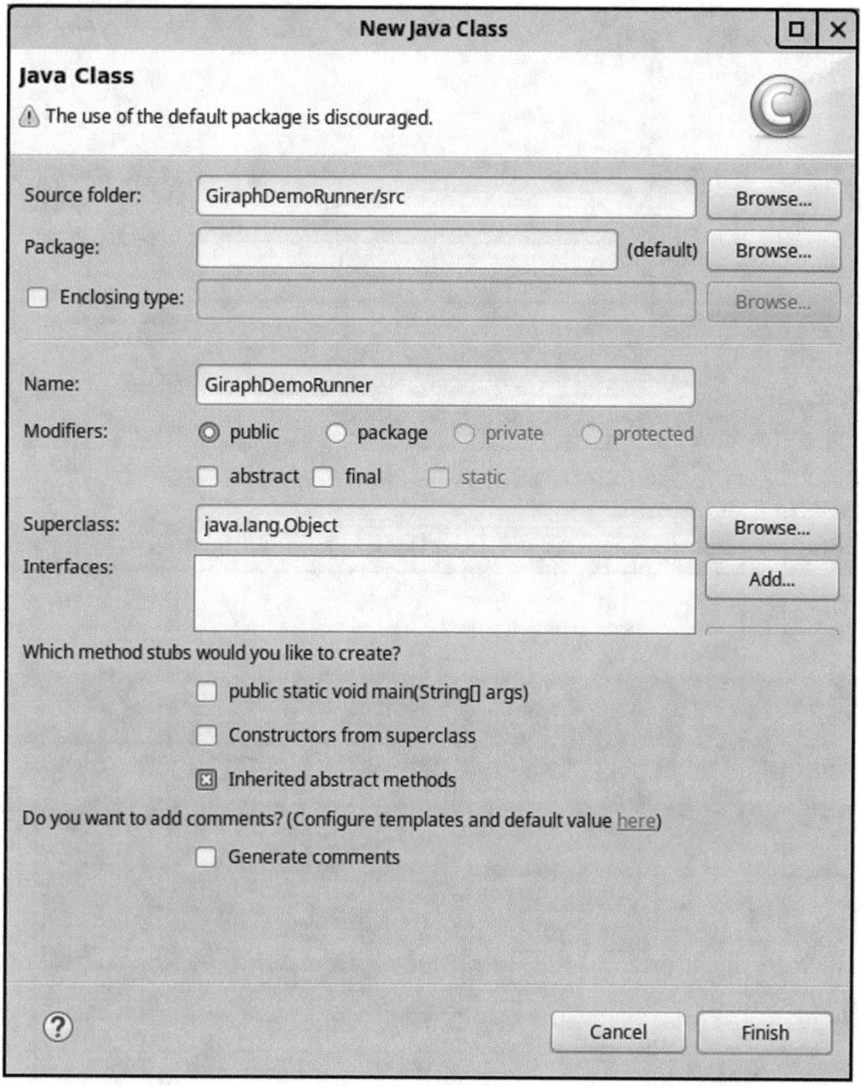

**Fig. 2.21**  Eclipse: new Java class window

6. A new window **Attach Files or Directories** (Fig. 2.29) will open in which you can choose the jar files or directories as dependencies to the project. Select the following jar files and directories:

   a.  hadoop-1.2.1/hadoop-core-1.2.1.jar
   b.  hadoop-1.2.1/hadoop-client-1.2.1.jar
   c.  hadoop-1.2.1/lib

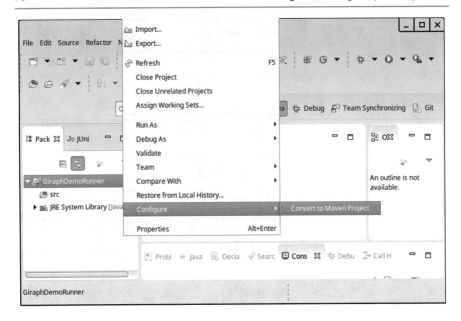

**Fig. 2.22** Eclipse: conversion to Maven project

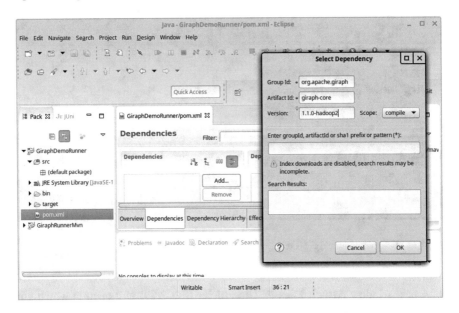

**Fig. 2.23** Eclipse: dependencies addition

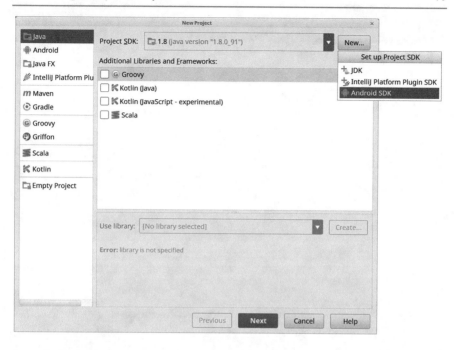

**Fig. 2.24** IntelliJ IDEA: new project window with JDK selection

    d.  giraph-1.1.0-for-hadoop-1.2.1/giraph-core-1.1.0.jar
    e.  giraph-1.1.0-for-hadoop-1.2.1/giraph-examples-1.1.0.jar
    f.  giraph-1.1.0-for-hadoop-1.2.1/lib

7.  Follow the steps from 7 to 10 in Sect. 2.9.1.

## 2.9.4   IntelliJ IDE with Maven

Follow the steps 1 to 4 in Sect. 2.9.3 to create a new project and name it `GiraphDemo RunnerMaven`.

1.  Right-click on the project name and choose **Add Framework Support ...** (Fig. 2.30).
2.  In the resulting window, check **Maven** and click on **OK** (Fig. 2.31).
3.  A `pom.xml` file will be created in the project's root directory and open it in the editor. On the top right corner of the IDE window (Fig. 2.32), you will find a tool-tip. Click on **Enable Auto-Import**. If the tool-tip does not appear, press `ctrl+S` to save the `pom.xml` file. You should see the tool-tip now.

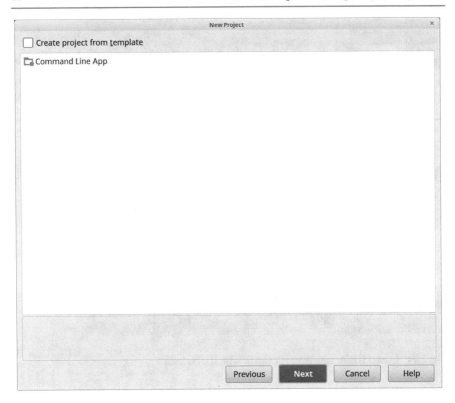

**Fig. 2.25** IntelliJ IDEA: new project template selection window

4. Add the following content of Listing 2.6 to the `pom.xml` file and save the file. After saving, IntelliJ will automatically download the dependencies.
5. Create a new class `GiraphDemoRunner` in `src/main/java` directory in the IntelliJ project window and add the content from Listing 2.4 to the newly created class file.
6. Create a `log4j.properties` file in the `src/main/resources` directory as described in step 5 of Sect. 2.9.2.
7. Now right-click on the file in the editor window and chose **run 'GiraphDemoRunner.main()'** to run the shortest path algorithm. You will see the console output in the 'Run' window at the bottom.
8. To check the output, open `/tmp/graph_out/part-m-00000` file and you should see something like:

```
0          1.0
1          0.0
2          2.0
3          1.0
4          5.0
```

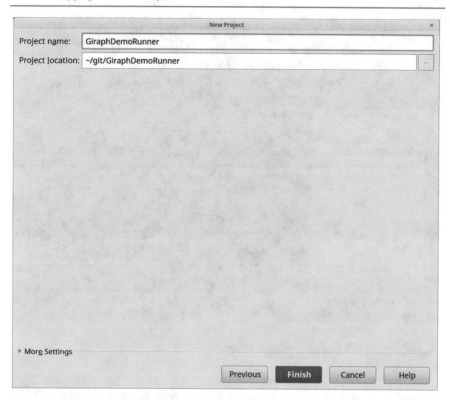

**Fig. 2.26** IntelliJ IDEA: new project name window

## 2.10   Debugging Local Zookeeper

While running a Giraph program inside an IDE, you might get the following exception multiple times:

```
WARN zk.ZooKeeperManager: onlineZooKeeperServers:
    Got SocketTimeoutException java.net.
    SocketTimeoutException: connect timed out
```

In the local mode, when Giraph does not find a running Zookeeper, it creates a local instance of Zookeeper itself. It creates a directory _bsp inside the project directory to store the data and the configuration for the zookeeper instance. If after creation of a Zookeeper instance, Giraph still can't connect to it, you can debug the problem using the following steps on a Linux OS (you can replicate similar steps for other OS).

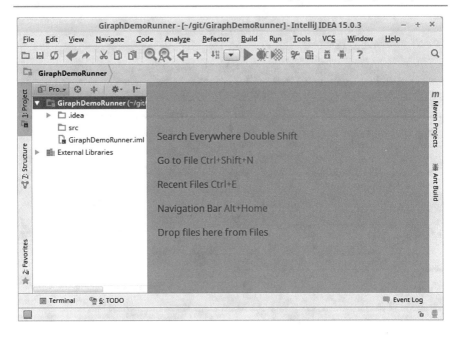

**Fig. 2.27**   IntelliJ IDEA: giraph project structure

1. Navigate to the project directory using the command line or a file browser. Open the file `_bsp/_defaultZkManagerDir/zoo.cfg`. It should look similar to the following:

```
tickTime=6000
dataDir=/home/faisal/workspace/GiraphDemoRunner/
    _bsp/_defaultZkManagerDir
clientPort=22181
maxClientCnxns=10000
minSessionTimeout=600000
maxSessionTimeout=900000
initLimit=10
syncLimit=5
snapCount=50000
forceSync=no
skipACL=yes
```

Note down the `clientPort` which in this case is `22181`.

2. By default, Giraph tries to connect to Zookeeper 10 times after which it fails. While Giraph is trying to connect to Zookeeper, type the following in the terminal:

```
lsof -i :22181
```

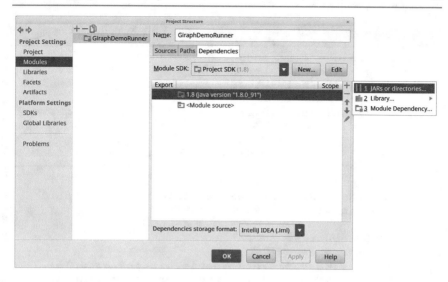

**Fig. 2.28** IntelliJ IDEA: project structure window

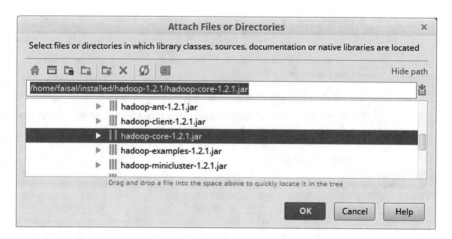

**Fig. 2.29** IntelliJ IDEA: attaching Jar files

The above command will list the processes listening on port `22181`. The output of this command should be similar to:

```
COMMAND PID USER FD TYPE DEVICE SIZE/OFF NODE
    NAME
java 3145 hadoop 188u IPv6 1757146 0t0 TCP
    *:22181(LISTEN)
```

---

**Listing 2.6: IntelliJ Maven Project pom.xml file**

```xml
<?xml version="1.0" encoding="UTF-8"?>
<project xmlns="http://maven.apache.org/POM/4.0.0"
         xmlns:xsi="http://www.w3.org/2001/XMLSchema
            -instance"
         xsi:schemaLocation="http://maven.apache.org
            /POM/4.0.0 http://maven.apache.org/xsd/
            maven-4.0.0.xsd">
    <modelVersion>4.0.0</modelVersion>

    <groupId>groupId</groupId>
    <artifactId>GiraphDemoRunnerMaven</artifactId>
    <version>1.0-SNAPSHOT</version>

    <dependencies>
        <dependency>
            <groupId>org.apache.giraph</groupId>
            <artifactId>giraph-core</artifactId>
            <version>1.1.0</version>
        </dependency>
        <dependency>
            <groupId>org.apache.hadoop</groupId>
            <artifactId>hadoop-core</artifactId>
            <version>1.2.1</version>
        </dependency>
        <dependency>
            <groupId>org.apache.giraph</groupId>
            <artifactId>giraph-examples</artifactId>
            <version>1.1.0</version>
        </dependency>
    </dependencies>
</project>
```

It shows that a Java process with process Id 3145 under user hadoop is listening on this port. If after the execution of the code is completed, the above command still shows a process listening on the port, this is probably another process accessing this port. You should either kill the process or change the default port of Zookeeper.

3. If the Zookeeper instance is running but Giraph cannot connect to it, most likely the problem is with the host naming. Following is a piece of the log Giraph puts on the console while trying to connect to the Zookeeper:

```
INFO zk.ZooKeeperManager: onlineZooKeeperServers
   : Connect attempt 9 of 10 max trying to
   connect to hadoopnode01:22181 with poll msecs
   = 3000
```

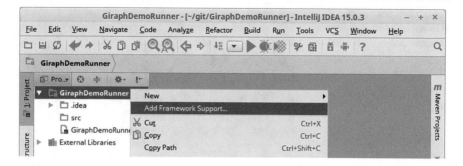

**Fig. 2.30** Add framework support to Intellij Java project

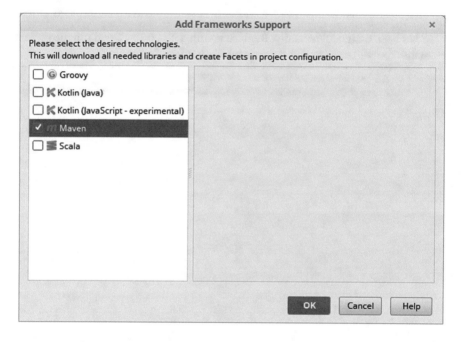

**Fig. 2.31** Intellij: Add Maven support to the project

We see that Giraph is trying to connect to hadoopnode01 which is the hostname of the machine we are using for running the Giraph project.

4. Ping the hostname:

```
ping hadoopnode01
```

If you are unable to ping the hostname, open the /etc/hosts file and note the ip address corresponding to the hostname. If it is other than 127.0.0.1, the

**Fig. 2.32**  Intellij: enable
auto-import of project
dependencies

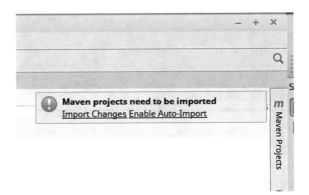

ip has been assigned manually. Replace it with `127.0.0.1` and run the project
again. You will have to edit the `hosts` file with `sudo` privileges:

```
sudo vim /etc/hosts
```

# Getting Started with Giraph Programming

**3**

## 3.1  Giraph Graph Model

Center to Apache Giraph graph model are the `Vertex` and the `Edge` interfaces. A graph is constructed using only these two interfaces. Figure 3.1 shows an `Edge`. An edge contains a value/weight and an id of the target vertex. In Giraph's graph model, a vertex contains an ID, a value, and a list of all its outgoing edges as shown in Fig. 3.2. Giraph provides default implementations of these interfaces, i.e., `DefaultVertex` and `DefaultEdge`. There are other implementations of these interfaces as well but in this section, instead of discussing those, we will be discussing the interfaces because while using the Giraph API, you will rarely touch the implementations.

### 3.1.1  The `Edge` Interface

Listing 3.1 shows the `Edge` interface. Type variables for the `Edge` interface are shown in Fig. 3.1. An edge contains a destination vertex ID of type variable `I` which extends `WritableComparable` and edge data of type variable `E` extending `Writable`. The types `I` and `E` depend on your graph type. The `InputFormat` chosen/implemented to load the graph data should comply with the type variable of edge. Note that `I` should be the same type as used in the `Vertex` as both represent a vertex id. Following are the two methods of the `Edge` interface.

Listing 3.1: Edge.java

```
public interface Edge<I extends WritableComparable, E extends
    Writable> {
  I getTargetVertexId();
  E getValue();
}
```

© Springer International Publishing AG 2016
S. Sakr et al., *Large-Scale Graph Processing Using Apache Giraph*,
DOI 10.1007/978-3-319-47431-1_3

**Fig. 3.1** `Edge <I`
`extends`
`WritableComparable,`
`E extends Writable>`

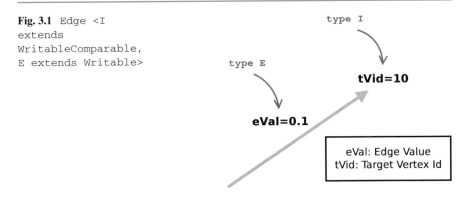

| |
|---|
| **getTargetVertexId()** |
| Get the target vertex ID of the vertex to which this edge points. |
| **getValue()** |
| Get the value of the edge. |

### 3.1.2  The `Vertex` Interface

Figure 3.2 shows a Giraph vertex. It contains a vertex id, vertex value, and a list of its out edges. Listing 3.2 shows the `Vertex` interface. It accepts a vertex ID of type variable `I`, vertex value of type variable `V`, and an `Iterable` of `Edge` class. As explained in the previous section, the `Edge` contains a target vertex ID of type `I` and an edge value of type `E`. The types `I`, `V`, and `E` depend on your graph type and a matching `InputFormat` should be chosen/implemented to load the graph data. In the following we explain the methods of the `Vertex` interface.

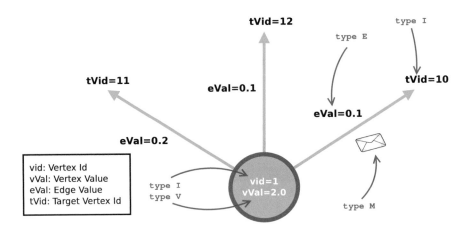

**Fig. 3.2** `Vertex <I extends WritableComparable, V extends Writable, E extends Writable>`

Listing 3.2: Vertex.java

```
public interface Vertex<I extends WritableComparable,
        V extends Writable, E extends Writable> extends
        ImmutableClassesGiraphConfigurable<I, V, E> {

    void initialize(I id, V value, Iterable<Edge<I, E>> edges);
    void initialize(I id, V value);
    I getId();
    V getValue();
    void setValue(V value);
    void voteToHalt();
    int getNumEdges();
    Iterable<Edge<I, E>> getEdges();
    void setEdges(Iterable<Edge<I, E>> edges);
    Iterable<MutableEdge<I, E>> getMutableEdges();
    E getEdgeValue(I targetVertexId);
    void setEdgeValue(I targetVertexId, E edgeValue);
    Iterable<E> getAllEdgeValues(final I targetVertexId);
    void addEdge(Edge<I, E> edge);
    void removeEdges(I targetVertexId);
    void unwrapMutableEdges();
    void wakeUp();
    boolean isHalted();
}
```

**I getId()**

This method returns the ID of the vertex.

**V getValue()** and void setValue(V value)

These methods return the vertex value (data stored with vertex) and set the vertex value respectively.

**void setEdges(Iterable<Edge<I, E>> edges)**

It takes as input an Iterable of edges and sets it for the vertex.

**int getNumEdges()**

Get the number of outgoing edges on this vertex

**void addEdge(Edge<I, E> edge)**

This method accepts an edge as a parameter and adds it to the vertex

**void removeEdges(I targetVertexId)**

Removes all edges pointing to the given vertex id.

**void wakeUp()**

This method re-activates the vertex if halted.

**boolean isHalted()**

It returns true if the vertex is halted and false otherwise.

**void voteToHalt()**

This method is used to halt a vertex. After this method is called, the compute() code will no longer be called for this vertex unless a message is sent to it. Any message to the vertex wakes it up and Giraph starts calling the compute() method until this method is called again. The application finishes only when all vertexes vote to halt.

**void initialize(I id, V value, Iterable<Edge<I, E>> edges)**

This method is used to initialize id, value, and edges (out-going) of a vertex. This method (or the alternative form initialize(id, value)) must be called after instantiation, unless a vertex is being deserialized and `readFields()` is called. You won't need to initialize a vertex if you are using an existing `InputFormat` (A class which reads the input graph data from a data source and creates a graph in memory) provided by Giraph.

**void initialize(I id, V value)**

This method initializes the id and value of the vertex. Vertex edges will be empty. This method (or the alternative form initialize(id, value, edges)) must be called after instantiation, unless the vertex is being de-serialized and `readFields()` is called. You won't need to initialize a vertex if you are using an existing `InputFormat` (A class which reads the input graph data from a data source and creates a graph in memory) provided by Giraph.

**Iterable<Edge<I, E>> getEdges()**

Get a read-only view of the out-edges of this vertex. The sort order of the returned edges is determined by the subclass implementation. Note: Edge objects returned by this `iterable` may be invalidated as soon as the next element is requested. Thus, keeping a reference to an edge almost always leads to undesired behaviour. Accessing the edges with other methods (e.g., `addEdge()`) during iteration leads to undefined behaviour (More on this later).

**Iterable<MutableEdge<I , E>> getMutableEdges()**

This method gets an `iterable` of out-edges that can be modified in-place. You can change the current edge value or remove the current edge while iterating through the edges. Note: Accessing the edges with other methods (e.g., addEdge()) during iteration leads to undefined behaviour.

**E getEdgeValue(I targetVertexId)**

This method returns the value of the first edge with the given target vertex id, or null if there is no such edge. Note that edge value objects returned by this method may be invalidated by the next call. Thus, keeping a reference to an edge value almost always leads to undesired behaviour.

**void setEdgeValue(I targetVertexId, E edgeValue)**

If an edge to the target vertex exists, this method sets it to the given edge value. This only makes sense with strict graphs (a graph with no parallel edges and self loops). If there are more than one edges directed to a vertex, you can only set the value of the first edge in the list.

**Iterable<E> getAllEdgeValues(final I targetVertexId)**

This method returns an `Iterable` over the values of all edges with the given target vertex id. This only makes sense for multigraphs (i.e. graphs with parallel edges). Note: Edge value objects returned by this method may be invalidated as soon as the next element is requested. Thus, keeping a reference to an edge value almost always leads to undesired behaviour.

### 3.1.3 The `Computation` Interface

The `Computation` interface represents the computation to be applied to all the alive vertexes in each superstep. During the superstep there can be several instances of this interface, each doing computation on one partition of the graph's vertexes. Note that each thread will have its own `Computation`, so accessing any data from this class is thread-safe. However, accessing global data (like data from `WorkerContext`) is not thread-safe. Objects of this interface only live for a single superstep. The most important method and the only one that is mandatory to write is the `compute()` method. The use of its other methods depends on the the type of algorithm you are implementing. In the following we describe the methods of this interface.

---

**`void compute(Vertex<I, V, E> vertex, Iterable<M1> messages)`**
This is the most important method of the computation interface. It is called one time on each alive vertex in each superstep. Remember that all those vertexes which have voted for a halt wake up when any other vertex has sent a message to it in the previous superstep. The `vertex` variable represents the target vertex whereas `messages` variable represent the messages that were sent to this vertex in the previous superstep. Each message is only guaranteed to have a life expectancy as long as next() is not called. If you want to use more than one message at a time, make a copy of the messages.

**`void preSuperstep()`**
This method is executed exactly once prior to compute(Vertex, Iterable) being called for any of the vertexes in the partition. It can be used to prepare for the superstep. Note that there is one computation interface per partition responsible for all computations in that partition hence, `preSuperstep()` should not be used for execution of the code required at the global level. You can use `MasterCompute` class for running global code just before a superstep.

**`void postSuperstep()`**
This method is executed exactly once after computation for all vertexes in the partition is complete. It is generally used for the code required to finish a superstep.

---

The following methods of the `Computation` interface are generally not required to be implemented as Giraph comes with a class `AbstractComputation` which has this method already implemented. You just have to extend this class for writing your computation after which you can call these methods whenever needed.

**`long getSuperstep()`**
Retrieves the current superstep.

**`long getTotalNumvertexes()`**
Gets the total number of vertexes in all the partitions that existed in the previous superstep. If it is the first superstep, it will return $-1$.

**`long getTotalNumEdges()`**
Gets the total number of edges in all partitions that existed in the previous superstep. If it is the first superstep, it will return $-1$.

**`void sendMessage(I id, M2 message)`**
Sends a message to a vertex. The parameter `id` represents the Vertex id to send the message to whereas the parameter `message` is the message data to send.

**`void sendMessageToAllEdges(Vertex<I, V, E> vertex, M2 message)`**
Send a message to all edges. The parameter `vertex` is the vertex whose edges to send the message to whereas the parameter `message` is the message sent to all the edges.

**`void sendMessageToMultipleEdges(Iterator<I> vertexIdIterator,`**
**`M2 message)`**
Sends the `message` to multiple target vertex ids in the iterator `vertexIdIterator`.

**`void addVertexRequest(I id, V value, OutEdges<I, E> edges)`**
**`throws IOException`**
Sends a request to create a vertex that will be available during the next superstep. The parameter `id` represents the vertex id, parameter `value` represents the vertex value and parameter `edges` represents the initial out-edges of the new vertex to be created.

**`void addVertexRequest(I id, V value) throws IOException`**
Sends a request to create a vertex with the given `id` and `value` that will be available during the next superstep. No edges of the vertex will be created.

**`void removeVertexRequest(I vertexId) throws IOException`**
Requests to remove a vertex with the given id from the graph. This mutation is applied just prior to the next superstep.

**`void addEdgeRequest(I sourceVertexId, Edge<I, E> edge) throws`**
**`IOException`**
Requests to add the given edge to a vertex with id `sourceVertexId` in the graph. This request is processed just prior to the next superstep. This method is different from the `addEdge()` method of the `Vertex` interface which adds the edge to the vertex immediately.

**`void removeEdgesRequest (I sourceVertexId, I targetVertexId)`**
Requests to remove all edges from the given source vertex to the given target vertex. The request is processed just prior to the next superstep.

**`Mapper.Context getContext()`**
Gets the mapper context. As Giraph runs on Hadoop, more specifically inside a Mapper, mapper context can be used to get the context information about the Mapper.

**`<W extends WorkerContext> W getWorkerContext()`**
Gets the worker context. `WorkerContext` contains context information about the Giraph worker responsible for execution of the current computation.

## 3.2  Vertex Similarity Algorithm

Let us use the knowledge gained in the previous sections to write a simple vertex similarity algorithm for finding friends' similarity in terms of their common friends. Figure 3.3 shows a simple social graph. Each vertex in the figure represents a person and the edges represent friendship. We want to assign a weight to each edge, representing the similarity between the persons it connects together. The weight of an edge will be higher if the persons it connects, have more common friends. Thus the weight an edge between two persons shows the similarity between the two persons. We use Jaccard Index as the similarity measure which is used for finding similarity between two sets. If A and B are two sets, Jaccard Index can be written as the size of the intersection between A & B divided by the size of the union of A & B:

$$J(A, B) = \frac{|A \cap B|}{|A \cup B|} = \frac{|A \cap B|}{|A| + |B| - |A \cap B|}$$

For instance, to calculate similarity between Peter and Jack, we can divide the number of common friends between them with the total number of their friends:

$$J(Jack, Peter) = \frac{|\{Peter, Eva\} \cap \{Jack, Tom, Eva\}|}{|\{Peter, Eva\} \cup \{Jack, Tom, Eva\}|}$$
$$= \frac{|\{Eva\}|}{|\{Peter, Jack, Tom, Eva\}|} = \frac{1}{4}$$

Vertex similarity can be calculated in two supersteps:

**Fig. 3.3**  Social graph

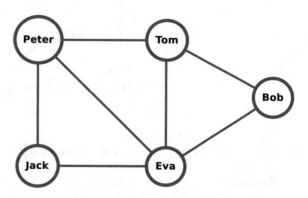

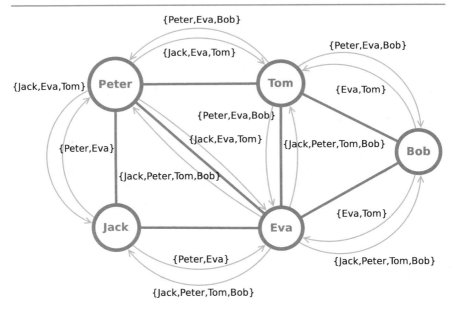

**Fig. 3.4**  Vertex similarity: superstep 0

**Fig. 3.5**  Vertex similarity: superstep 1

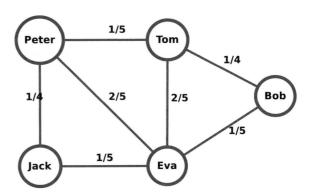

1. **Superstep 0**: In the first superstep, each person (vertex) sends the list of his/her friends to each of his/her friends.
2. **Superstep 1**: Each person (vertex) receives the friends's list of his/her friends and calculates its Jaccard similarity with the corresponding friend (Figs. 3.4 and 3.5).

## 3.2.1  Implementation in Giraph

Listing 3.3 shows the MasterCompute class whose compute() method is called before every superstep by Giraph. In this method, we set different computation classes for superstep 0 and 1. For superstep 0, we set the computation class to SendFriendsList.class which when executed for each vertex, sends the vertex's list of neighbors to all its neighbors. For superstep 1, we set the computation class to JaccardComputation.class.

Listing 3.3: MasterCompute.java

```java
public static class MasterCompute extends
    DefaultMasterCompute {

    @Override
    public final void compute() {
        long superstep = getSuperstep();
        if (superstep == 0) {
            setComputation(SendFriendsList.class);
        } else if (superstep == 1) {
            setComputation(JaccardComputation.class);
        }
    }
}
```

Listing 3.4 shows the SendFriendsList computation class which will be executed for each vertex in superstep 0. The compute() method accepts the vertex for which the computation should be executed and all the messages sent to it in the previous superstep. As we are at superstep 0, we will not have any messages yet. Remember that the purpose of this computation is to send the vertex's list of friends (neighbors) to all its neighbors. We first create an empty friend list (line 10). We then parse each edge of the vertex (line 12), get the id of the vertex it points to (line 13), and fill it in the friend list we created before (line 14). It is important to clone the vertex id before adding it to the list as Giraph provides a temporary reference to all edges, in the computation method. After creating the friends list, we create a message object LongIdFriendsList (line 17), set the current vertex id as the source of the message (line 19), fill the message with the friends list (line 20), and send the message to all its neighbors (line 21).

Listing 3.4: SendFriendsList.java

```
1  public static class SendFriendsList
2          extends BasicComputation<LongWritable,
3          NullWritable, DoubleWritable,
              LongIdFriendsList> {
4
5      @Override
6      public void compute(Vertex<LongWritable,
          NullWritable, DoubleWritable> vertex,
7                      Iterable<LongIdFriendsList>
                          messages)
8              throws IOException {
9
10         final LongArrayListWritable friends =  new
              LongArrayListWritable();
11
12         for (Edge edge : vertex.getEdges()) {
13             LongWritable id = (LongWritable)edge.
                  getTargetVertexId();
14             friends.add(WritableUtils.clone(id, getConf
                  ()));
15         }
16
17         LongIdFriendsList msg = new LongIdFriendsList()
              ;
18
19         msg.setSourceId(vertex.getId());
20         msg.setMessage(friends);
21         sendMessageToAllEdges(vertex, msg);
22     }
23  }
```

In the second superstep (superstep 1), the computation class in Listing 3.5 is applied to all vertexes. The `compute()` method receives a vertex and all the messages (friends list) sent to it by its neighbors in the previous superstep. Each message contains a friends list from one of its neighbors. The messages are traversed one by one (line 10). For a message we find its source (line 11) and try to check if the current vertex has an edge with the sender of the message. This is done by retrieving the edge value between the current vertex and the message sender (line 12). If the returned edge value is `null`, it means that the sender is not the vertex's neighbor and an error is thrown (line 13). We traverse each vertex id in the friends list contained in the received message (line 16) and check if the current vertex also has it as a neighbor (line 17). If it is, then the vertex is a common neighbor (line 18). After the traversal of the vertex ids contained in the message, the Jaccard index is calculated and is set as the value of the edge between the message sender and the current vertex (line 24). This is repeated for rest of the received messages. In the end, the vertex votes for halt (line 27) telling Giraph that it does not have any more work to do.

Listing 3.5: JaccardComputation.java

```
 1  public static class JaccardComputation extends
        BasicComputation<LongWritable,
 2            NullWritable, DoubleWritable,
                LongIdFriendsList> {
 3
 4
 5        @Override
 6        public void compute(
 7                Vertex<LongWritable, NullWritable,
                    DoubleWritable> vertex,
 8                Iterable<LongIdFriendsList>
                    messages) throws IOException {
 9
10            for (LongIdFriendsList msg : messages)
                {
11                LongWritable src = msg.getSourceId
                    ();
12                DoubleWritable edgeValue = vertex.
                    getEdgeValue(src);
13                assert(edgeValue!=null);
14                long totalFriends = vertex.
                    getNumEdges();
15                long commonFriends = 0;
16                for (LongWritable id : msg.
                    getMessage()) {
17                    if (vertex.getEdgeValue(id)!=
                        null) {
18                        commonFriends++;
19                    } else {
20                        totalFriends++;
21                    }
22                }
23
24                vertex.setEdgeValue(src, new
                    DoubleWritable(
25                        (double)commonFriends/(
                            double)totalFriends));
26            }
27            vertex.voteToHalt();
28        }
29    }
```

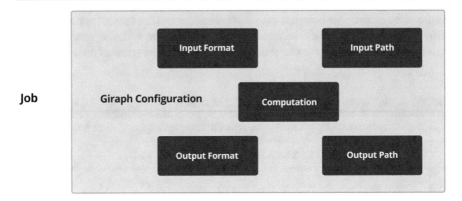

**Fig. 3.6** Basic job composition

After the second superstep (superstep 1), all vertexes will be halted and Giraph will end the job after outputting the current graph to the disk.[1]

## 3.3  Writing a Basic Giraph Job

Now that you know about the three most important Giraph interfaces, you can start constructing a Giraph job. In this section, we assume that the computation class has already been implemented. If you are reading this section without following the previous sections about writing the computation class, do not worry as we will use the shortest path example that comes packaged with Giraph's examples jar. Figure 3.6 shows the minimal composition of a Giraph Job. A Giraph job is created by passing it a `GiraphConfiguration` object which contains different properties which Giraph uses to execute the job. The primary property is the `Computation` class which contains the `compute()` method. This compute method is applied to all alive vertexes in each superstep. Giraph also needs to know where the input data is located and how to read that data and create a graph. Giraph uses the `InputFormat` class specified in the `GiraphConfiguration` object to read the input data and create a graph in memory. When the supersteps are finished, it uses the `OutputFormat` class to convert the resultant graph in memory to a writable format and stores it in the directory specified as the output path in the configuration object. There are other properties as well which can be modified to customize Giraph's execution behavior but if they are not specified, the defaults are used by Giraph.

---

[1]The source code for this example can be found at https://github.com/sakrsherif/GiraphBookSourceCodes/tree/master/chapter03.

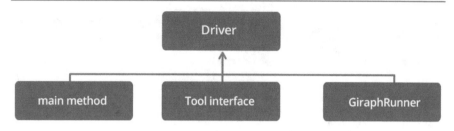

**Fig. 3.7** Ways of writing a driver program

Listing 3.6 shows a job for the shortest path algorithm. It can be seen that the code creates a `GiraphConfiguration` object and fills it in with the five compulsory components of the job. We use the `SimpleShortestPathsComputation` which comes packaged in the `giraph-examples-*.jar`. The property `setLocalTestMode(true)` tells Giraph to run the job locally on the machine using threads. The local mode produces the same job output as in distributed mode, however it simulates multiple workers using different threads. Local mode is generally used for development and debugging purposes. The `SPLIT_MASTER_WORKER` property tells Giraph to run the master and worker tasks separately. Setting this property to `true` will result in an error because in local mode, the job is executed by the `LocalJobRunner` which fires up only one worker to execute the job. In production environment, you should set this property to `true`. The `setWorkerConfiguration()` sets the minimum and maximum number of workers to fire up for the job. In local mode, not setting this property or setting it more than 1 causes an error.

## 3.4   Writing the Driver Program

All Giraph jobs are created and sent for execution from the driver program. Generally a driver program runs on the master and submits the job to the `JobTracker` for execution. Creating a job is the job of a driver program. Figure 3.7 shows three different ways of writing a driver program which are explained in the following.[2]

---

[2]The source code for the driver writing methods can be found at https://github.com/sakrsherif/ GiraphBookSourceCodes/tree/master/chapter03.

**Listing 3.6: GiraphDriverMain.java**

```
import org.apache.giraph.conf.GiraphConfiguration;
import org.apache.giraph.examples.
    SimpleShortestPathsComputation;
import org.apache.giraph.io.formats.GiraphFileInputFormat;
import org.apache.giraph.io.formats.
    IdWithValueTextOutputFormat;
import org.apache.giraph.io.formats.
    JsonLongDoubleFloatDoubleVertexInputFormat;
import org.apache.giraph.job.GiraphJob;
import org.apache.hadoop.fs.Path;
import org.apache.hadoop.mapreduce.lib.output.
    FileOutputFormat;


public class GiraphDriverMain {
    public static void main(String[] args) throws
        Exception {
        String inputPath = "/tmp/input.txt";
        String outputPath = "/tmp/output";

        //create GiraphConfiguration object
        GiraphConfiguration giraphConf = new
            GiraphConfiguration();
        //set Computation class
        giraphConf.setComputationClass(
                SimpleShortestPathsComputation.class);
        //set InputFormat class
        giraphConf.setVertexInputFormatClass(
                JsonLongDoubleFloatDoubleVertexInputFormat
                    .class);
        //set Input Path
        GiraphFileInputFormat.addVertexInputPath(
                giraphConf, new Path(inputPath));
        //set other properties
        giraphConf.setLocalTestMode(true);
        giraphConf.SPLIT_MASTER_WORKER.set(
                giraphConf, false);
        giraphConf.setWorkerConfiguration(1, 1, 100);
        //set OutputFormat class
        giraphConf.setVertexOutputFormatClass(
                IdWithValueTextOutputFormat.class);
        //create Giraph Job object
        GiraphJob giraphJob = new GiraphJob(
                giraphConf, "GiraphDriverMain");
        //set Output Path
        FileOutputFormat.setOutputPath(
                giraphJob.getInternalJob(),
                new Path(outputPath));
        //run Giraph Job
        giraphJob.run(true);
    }
}
```

### 3.4.1   Using `main` Method

One way to write the driver program is to construct the `GiraphConfiguration` object in the `main()` method, specify the required properties, e.g., computation class, IO formats, etc, create a `GiraphJob` object using the `Giraph Configuration` object and submit it for execution. This method is good for initial development and testing but is not flexible and distributable because the Giraph properties/options are fixed in the code. After a JAR file had been created for distribution, if any change is required, the code needs to be changed and a new JAR is created. Listing 3.6 shows a driver program which constructs the Giraph job in the `main()` method.

One way to fix this problem is to accept the changeable properties as arguments to the `main()` method and set them in the `GiraphConfiguration` object. This solution is a bit better than fixing the property values in the code but it is not much flexible. There are a number of properties/options that Giraph supports. Writing code for accepting all properties and their values is a tedious task. The good news is that Hadoop provides a `ToolRunner` class which takes care of this problem.

### 3.4.2   Using `Tool` Interface

Hadoop comes with a helper class `ToolRunner` which is capable of parsing the command line arguments and creating a Hadoop `Configuration` object which you can use to create a `GiraphConfiguration` object. `ToolRunner` can run classes implementing the `Tool` interface. It works in conjunction with `GenericOptionsParser` to parse the generic `hadoop` command line arguments and modifies the `Configuration` of the Tool. The application-specific options are passed along without being modified. The driver program should implement the `Tool` interface and should have a `Configuration` object. `ToolRunner` will parse the command line arguments and set the properties in the `Configuration` object. In the `run()` method of the driver class, the `GiraphJob` should be constructed using the `Configuration` object and submitted.

Listing 3.7 shows a basic driver class using the `Tool` interface. It can be seen that all the properties except for the `Computation` class and the job name have been moved out of the code. These properties can now be provided to the driver through the command line. If you want to run `GiraphDriverTool.java` inside an IDE, you should provide the arguments in the run configuration of the class as shown in the Listing 3.8.

The "`-D`" argument indicates the start of a property followed by its value. All the properties are automatically parsed by the `ToolRunner` class and put in the job configuration. In the current form, we can distribute the Java file to other users who can choose the properties, e.g., `InputFormat` based on their type of data.

**Listing 3.7: GiraphDriverTool.java**

```java
import org.apache.giraph.conf.GiraphConfiguration;
import org.apache.giraph.examples.
    SimpleShortestPathsComputation;
import org.apache.giraph.job.GiraphJob;
import org.apache.hadoop.conf.Configuration;
import org.apache.hadoop.util.Tool;
import org.apache.hadoop.util.ToolRunner;


public class GiraphDriverTool implements Tool{
    private Configuration conf;
    public Configuration getConf() {
        return conf;
    }

    public void setConf(Configuration conf) {
        this.conf = conf;
    }

    public int run(String[] args) throws Exception
        {
        GiraphConfiguration giraphConf = new
            GiraphConfiguration(getConf());
        giraphConf.setComputationClass(
            SimpleShortestPathsComputation.class);
        GiraphJob giraphJob = new GiraphJob(
            giraphConf,
                "GiraphDriverTool");
        giraphJob.run(true);
        return 0;
    }

    public static void main(String[] args) throws
        Exception{
        ToolRunner.run(new GiraphDriverTool(), args
            );
    }
}
```

Listing 3.8: Arguments for Giraph Driver implementing Tool interface

```
-D giraph.vertexInputFormatClass=org.apache.
   giraph.io.formats.
   JsonLongDoubleFloatDoubleVertexInputFormat \
-D giraph.vertex.input.dir="/home/User/git/
   GiraphDemoRunnerMaven/src/main/resources/input
   .txt" \
-D giraph.minWorkers=1 \
-D giraph.maxWorkers=1 \
-D giraph.SplitMasterWorker=false \
-D giraph.localTestMode=true \
-D mapred.output.dir="/home/User/git/
   GiraphDemoRunnerMaven/src/main/resources/
   output" \
-D giraph.vertexOutputFormatClass="org.apache.
   giraph.io.formats.IdWithValueTextOutputFormat"
```

### 3.4.3   Using `GiraphRunner` Class

On the one hand `Tool` interface helps in making the Giraph job flexible, on the other hand it makes it difficult to run the job as the user has to enter long property names on the command line, e.g., `giraph.vertexInputFormatClass`. Giraph comes with a `Tool` called `GiraphRunner`. It makes it easier to write frequently used Giraph property names, e.g., instead of writing `giraph.vertexInputFormat Class` to specify the vertex input format, we can just write -vif. Listing 3.9 shows how to use `GiraphRunner` to write a driver program. Listing 3.10 shows the arguments for the driver. The first argument is the `Computation` class which is mentioned without any parameter specifier. The parameter specifier -vif denotes the property `giraph.vertexInputFormatClass`. Note that all the frequently used property names have been replaced by their intuitive abbreviations. The properties for which the `GiraphRunner` does not support abbreviations can still be specified using the parameter specifier -ca which stands for "custom argument." All properties including Hadoop configuration properties specified after the -ca parameter specifier are added to the Job Configuration object. The properties for which `GiraphRunner` supports dedicated parameter specifiers can be found by running the following command:

```
giraph org.apache.giraph.GiraphRunner -h
```

Properties specific to the algorithm can still be specified in the code. Line 9 in Listing 3.9 specifies the maximum number of supersteps for the algorithm. Any properties set inside the code override those provided at the command line. `GiraphRunner` is a tool which can be used as a driver without writing our own driver program. If there are no algorithm-specific properties that you want to fix in the code, writing a

**Listing 3.9: GiraphDriverGiraphRunner.java**

```
 1  import org.apache.giraph.GiraphRunner;
 2  import org.apache.giraph.conf.GiraphConfiguration
       ;
 3  import org.apache.hadoop.util.ToolRunner;
 4
 5  public class GiraphDriverGiraphRunner {
 6      public static void main(String[] args) throws
            Exception {
 7          GiraphRunner giraphRunner = new
                GiraphRunner();
 8          giraphRunner.setConf(new
                GiraphConfiguration());
 9          ((GiraphConfiguration)giraphRunner.
                getConf()).setMaxNumberOfSupersteps
                (100);
10          ToolRunner.run(giraphRunner, args);
11      }
12  }
```

driver program is not necessary as `GiraphRunner` can be used from the command line to construct a job. Section 2.5 shows how to use `GiraphRunner` to construct a job from the command line. Writing your own driver program is always helpful during the development phase because you can run and debug your code in the IDE.

**Listing 3.10: Arguments for Giraph Driver using GiraphRunner class**

```
org.apache.giraph.examples.
    SimpleShortestPathsComputation
-vif org.apache.giraph.io.formats.
    JsonLongDoubleFloatDoubleVertexInputFormat
-vip "/home/User/git/GiraphDemoRunnerMaven/src/
    main/resources/input.txt"
-vof org.apache.giraph.io.formats.
    IdWithValueTextOutputFormat
-op /home/User/git/GiraphDemoRunnerMaven/src/main
    /resources/output
-w 1
-ca giraph.SplitMasterWorker=false
-ca giraph.localTestMode=true
```

## 3.5  Preparing Graph Data for Giraph Input

### 3.5.1  Hadoop Distributed File System

Before preparing input data for Giraph, it is important to know how a Giraph job accesses its input data, which explains why it is not as simple as just uploading an edge list or a CSV file. Giraph runs on Hadoop and can access data from HDFS (Hadoop Distributed File System). HDFS is the preferred file system for running Giraph in the distributed mode. HDFS divides the data into chunks (called Blocks) and distributes them on different machines in a cluster. When a Giraph job is submitted to the cluster, "Workers" are launched on different machines of the cluster which execute the user written code on vertexes in parallel. These workers are controlled by a *Master* thread running on one of the machines. Based on the number of workers available, the Master logically divides the input data into chunks (called splits) and assign them to the workers to process. The workers connect with HDFS to get their part of the data, i.e., split. The split can be on the same machine where the worker is running or can be on a remote machine in which case it will be retrieved over the network. Using HDFS gives Giraph a number of benefits:

1. Giraph can process huge data sizes which cannot fit in the secondary storage of a single machine.
2. HDFS has the ability to replicate the data. If one machine containing the data fails, it is retrieved from another machine containing a copy of the data.
3. HDFS allows reading data in parallel. If a worker A has been assigned a split located on machine $x$ and a worker B has been assigned a split located on machine $y$, both the workers can connect to the machines holding their respective splits independently and in parallel. This results in a faster IO.
4. HDFS allows increased data locality which means that Giraph can assign splits to the workers which are located on the same machine as the workers thus reducing network IO.
5. HDFS provides one unified interface for data access independent of the execution location of Giraph. It is possible to have Giraph on a separate cluster different from that of the HDFS as long as there is network connectivity between them.

### 3.5.2  Hadoop Input Formats

HDFS was developed as part of the Hadoop project which uses it as its primary file system. It could be daunting for a novice user to write code for accessing data from HDFS and splitting it among different machines in the cluster. Therefore, Hadoop comes with different input formats out of the box. An input format is not more than a Java class which contains logic for accessing and splitting the data in HDFS. One can a write customized format and provide it to Hadoop to read data in a different way. Hadoop provides input formats for reading many different types including but not limited to text, binary, and sequence files. One famous input format is the `TextInputFormat` which reads text files from the HDFS line by line. If you

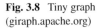
**Fig. 3.8** Tiny graph
(giraph.apache.org)

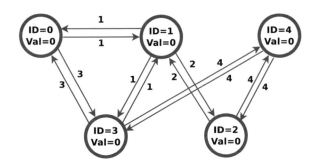

have a text file with graph data such that each line of the file contains a vertex and its associated data, you can extend the `TextInputFormat` class in a way that instead of returning a line of text, it returns a vertex object. That is exactly what Giraph does. It extends/modifies different Hadoop input formats so that they can return different graph elements such as vertexes, edges, etc. This way, you do not have to worry about how to connect to HDFS and get the data.

### 3.5.3  Giraph Input Formats

This section will show you how can you load your graph data using the rich set of input formats that come with Giraph. If none of the provided input formats fit your graph data format, you can either transform it before loading it to Giraph or write your own input format. Writing a custom input format is covered in later chapters of the book. To explain the different input formats, we use the graph `tiny_graph.txt` from **giraph.apache.org** shown in Fig. 3.8. It contains 5 vertexes and 12 edges. Note that a bidirectional edge is represented by two opposite unidirectional edges with the same weight. The reason for this is that Giraph does not support bidirectional edges.

Giraph represents a graph vertex with an ID of type `I` and a value of type `V`. An edge is represented by a source vertex ID, target vertex ID, and and edge weight of type `E`. `I`, `V`, and `E` can be any Java type as long as they implement the `Writable` interface which Hadoop and Giraph use for serialization and de-serialization when the data is shared between machines over the network. Hadoop provides its own implementation of Java primitive types which implement the `Writable` interface, e.g., `LongWritable`, `DoubleWritable`, `FloatWritable`, etc. If we want to have a different type for vertex ID, vertex value, or edge weight, we can implement the `Writable` interface for those types so that Giraph knows how to serialize and de-serialize them. Giraph supports three types of input formats:

1. Vertex Input Format
2. Edge Input Format
3. Mapping Input Format

We explain these types in the following sections.

### 3.5.4  Giraph Vertex Input Formats

VertexInputFormat is an abstract class which represents the category of input formats that can read vertexes. As the name implies, the input formats derived from VertexInputFormat can only read vertexes. If edge information is to be read, either it can be embedded in the corresponding vertexes or a separate EdgeInputFormat can be used. In Giraph's graph model, each vertex contains a list of all edges originating from that vertex. For representing bidirectional edges, we need to have two opposing unidirectional edges stored with the two corresponding vertexes. For example, in Fig. 3.8, the edge between vertex 0 and vertex 1 is a bidirectional edge. To represent that, vertex 0 will store a unidirectional edge directed toward vertex 1, whereas vertex 1 will store a unidirectional edge directed toward vertex 0. More details on representing undirected graphs in Giraph can be found in Sect. 5.2.2. The primary requirement for using a VertexInputFormat is that all the outgoing edges of a vertex should be stored with the vertex. This way, while reading the vertex, the input format can also read all its outgoing edges which it can embed in the vertex. To control the input format of a Giraph job, users have to modify the value of giraph.vertexInputFormatClass with their desired VertexInputFormat class. In the following section, we explore different vertex input formats that come packaged with Giraph.

#### 3.5.4.1  JsonLongDoubleFloatDoubleVertexInputFormat

This vertex input format is used to read graph stored in JSON format. It reads long vertex ID's, double vertex values and float out-edge weights, and double message types. The message type specifies the Java type of messages that the vertexes can exchange during superstep execution. The files should be in the following JSON format: JSONArray( , , JSONArray(JSONArray( , ), ...)). Each line in the input file should represent a vertex using the following format:

```
[vertexID,vertexVal, [[targetID,edgeWeight],[targetID,edgeWeight],...]
```

For using this input format, the graph of Fig. 3.8 should be represented as follows:

```
1  [0,0,[[1,1],[3,3]]]
2  [1,0,[[0,1],[2,2],[3,1]]]
3  [2,0,[[1,2],[4,4]]]
4  [3,0,[[0,3],[1,1],[4,4]]]
5  [4,0,[[3,4],[2,4]]]
```

The first line represents vertex with ID 0, its value 0, and its outgoing edge list containing two edges. The first edge has a target vertex ID 1 and an edge weight 1, whereas the second edge has a target ID 3 and an edge weight 3 as well. We can tell Giraph to use this input format by adding it to Giraph configuration object:

```
1  giraphConf.setVertexInputFormatClass(
2          JsonLongDoubleFloatDoubleVertexInputFormat.class);
```

### 3.5.4.2  IntIntNullTextVertexInputFormat

This input format is used for reading unweighted graphs with `int` ids stored in text files. The vertex IDs and values are read as `int`. Each line of the text file should contain a vertex ID, its value, and its neighbors, all separated with a tab (`'\t'`). The graph of Fig. 3.8 can be represented as follows:

```
1   0    0    1    3
2   1    0    0    2    3
3   2    0    1    4
4   3    0    0    1    4
5   4    0    3    2
```

Here, each line represents a single vertex. Line 3 represents a vertex with ID 2 and value 0, and edges directed toward vertexes 1 and 4. We can tell Giraph to use this input format by adding it to Giraph configuration object:

```
1   giraphConf.setVertexInputFormatClass(
2         IntIntNullTextVertexInputFormat.class);
```

### 3.5.4.3  LongDoubleDoubleTextInputFormat

This input format is similar to `IntIntNullTextVertexInputFormat` and used for loading unweighted graphs from text files with the difference that it loads vertex IDs as `long` and vertex values as `double`. After loading each vertex (with unweighted edges), it calculates and assigns the weight (of type `double`) to each edge using the formula (Fig. 3.9):

$$\text{weight} = \frac{1}{\text{no. of outgoing edges}}$$

Figure 3.10 shows how an input unweighted graph is converted to a weighted graph by `LongDoubleDoubleTextInputFormat`. We can tell Giraph to use this input format by adding it to Giraph configuration object:

```
1   giraphConf.setVertexInputFormatClass(
2         LongDoubleDoubleTextInputFormat.class);
```

### 3.5.4.4  IntIntNullTextInputFormat

This input format is used to read graphs having vertexes without values and edges without weights. Each line in the input file should represent a vertex using the following format:

```
vertexID    neighbour1ID    neighbour2ID    neighbour3ID...
```

Note that each value should be separated with a tab character ('\t'). For using this input format, the graph of Fig. 3.8 should be represented as follows:

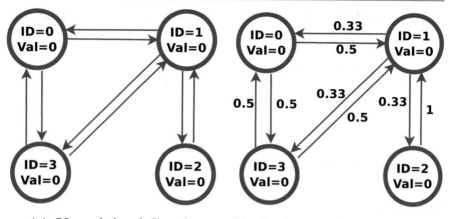

(a) Unweighted Graph          (b) Converted weighted graph

**Fig. 3.9**  Graph conversion by `LongDoubleDoubleTextInputFormat`

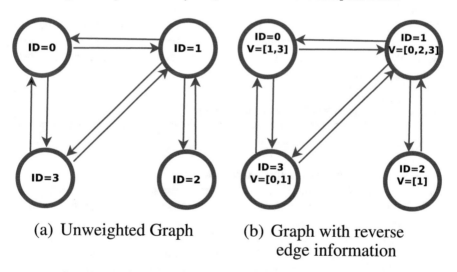

(a) Unweighted Graph          (b) Graph with reverse
                                  edge information

**Fig. 3.10**  Graph conversion by extttSccLongLongNullTextInputFormat

```
1  0   1   3
2  1   0   2   3
3  2   1   4
4  3   0   1   4
5  4   3   2
```

Here, each line represents a vertex. The first number is the vertex ID and the rest of
the numbers in a line are the vertex IDs of its neighbors to which it has an outgoing

edge. We can tell Giraph to use this input format by adding it to Giraph configuration object:

```
1 giraphConf.setVertexInputFormatClass(
2         IntIntNullTextInputFormat.class);
```

### 3.5.4.5  SccLongLongNullTextInputFormat

This input format is similar to `IntIntNullTextInputFormat` and is used to read unweighted graphs (without vertex values and edge weights) with two differences:

1. It reads vertex IDs as `long`
2. For a vertex value, it creates an instance of `SccVertexValue` class which can be used to hold neighbors which have edges toward this vertex.

Note that the input format itself does not add this information to the `SccVertex Value` object rather we need to add this information ourselves in the first superstep. This will be explained in later sections. It is also used in *Strongly Connected Component* algorithm. We can tell Giraph to use this input format by adding it to Giraph configuration object:

```
1 giraphConf.setVertexInputFormatClass(
2         SccLongLongNullTextInputFormat.class);
```

### 3.5.4.6  JsonBase64VertexInputFormat

This input format is a simple way to represent the structure of the graph with a JSON object. The actual vertex ids, values, edges are stored by the Writable serialized bytes that are Base64 encoded. It can also read graphs stored with `JsonBase64VertexOutputFormat`. The input file should contain one JSON object per line containing a vertex. The vertex ID, value, and edges should be stored under the keys provided by `JsonBase64VertexFormat` class which are:

1. `JsonBase64VertexFormat.VERTEX_ID_KEY = "vertexId"`
2. `JsonBase64VertexFormat.VERTEX_VALUE_KEY = "vertex Value"`
3. `JsonBase64VertexFormat.EDGE_ARRAY_KEY = "edgeArray"`

It creates an instance of `JSONObject` from the text on each line, extracts the strings corresponding to the above-mentioned keys, and decodes using `Base64` class. We can tell Giraph to use this input format by adding it to Giraph configuration object:

```
1  giraphConf.setVertexInputFormatClass(
2          JsonBase64VertexInputFormat.class);
```

### 3.5.4.7   IntNullNullTextInputFormat

This input format is used to load unweighted graphs with no edges. Each line of the
file should contain only a vertex ID. This input format can be used for applications
such as synthetic graph generators which load only vertex IDs initially and then
mutate the graph to create edges using statistical methods. We can tell Giraph to use
this input format by adding it to Giraph configuration object:

```
1  giraphConf.setVertexInputFormatClass(
2          IntNullNullTextInputFormat.class);
```

### 3.5.4.8   NormalizingLongDoubleDoubleTextInputFormat

It is a simple text-based input format for unweighted graphs with long ids but no
vertex values. Each line in the input file should contain a vertex with the format:

```
vertexID neighbor1:weight1 neighbor2:weight2 ...
```

Note that the edges should be separated with one or more spaces. The unweighted
graph of Fig. 3.11a can be represented as follows:

```
1  0    1:1  3:3
2  1    0:1  2:2  3:1
3  2    1:2  4:4
4  3    0:3  1:1  4:4
5  4    3:4  2:4
```

The normalized weight of an outgoing edge from a vertex $v$ is calculated using the
following formula:

$$\text{normalized weight} = \frac{\text{weight}}{\text{sum of weights of all outgoing edges of vertex } v}$$

### 3.5.4.9   LongLongNullTextInputFormat

This input format is similar to `IntNullNullTextInputFormat` and is used to
load unweighted graphs with `long` vertex IDs. It uses the vertex ID as the vertex
value as well.

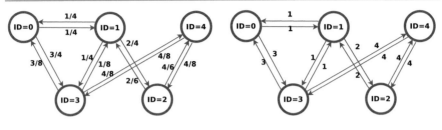

(a) Unweighted Graph            (b) Graph with normalized edges

**Fig. 3.11** Graph conversion by `NormalizingLongDoubleDoubleTextInputFormat`

### 3.5.4.10  LongDoubleFloatTextInputFormat

This input format is similar to `NormalizingLongDoubleDoubleTextInput Format` with a difference that it does not normalize edge weights. Instead it loads the same edge weights as specified in the input file.

### 3.5.4.11  LongDoubleDoubleAdjacencyListVertexInputFormat

This input format is used for reading graphs stored as (ordered) adjacency lists with the vertex ids of type `long` and the vertex values and edges of type `double`. Each line of the input file should contain a vertex with following format:

```
vertexID vertexValue neighbourID edgeValue neighbourID edgeValue ...
```

Note that each value in a line should be separated with a tab character (`| "\t"`). The tab character can be replaced as a separator by specifying a different character in `adj.list.input.delimiter` property of Giraph configuration. The graph of Fig. 3.8 can be represented in the input file as:

```
1  0  0  1  1  3  3
2  1  0  0  1  2  2  3  1
3  2  0  1  2  4  4
4  3  0  0  3  1  1  4  4
5  4  0  3  4  2  4
```

### 3.5.4.12  TextDoubleDoubleAdjacencyListVertexInputFormat

This input format is similar to `LongDoubleDoubleAdjacencyListVertex InputFormat` with a difference that it excepts vertex IDs as strings. This is a good input format for reading graphs where the id types do not matter and can be stashed in a `String`.

### 3.5.5   Edge Input Formats

Edge input formats are used to read graphs stored in the form of edges. When using edge input formats, it is not necessary to have all edges related to a vertex next to each other. The limitation of edge input formats is that they can not read vertex values. We explain each of the edge input formats that comes with Giraph but one can write a custom edge input format if required. Similar to `VertexInputFormat`, the `EdgeInputFormat` of a Giraph job is controlled through property `edgeInputFormatClass`.

#### 3.5.5.1   IntNullTextEdgeInputFormat

It is a simple text-based `EdgeInputFormat` for unweighted graphs with `int` ids. Each line of the input file should contain one edge using the format:

```
source_vertex_id, target_vertex_id
```

## 3.6   Preparing Graph Data for Giraph Output

Just like Giraph input formats, we can output the graph with one vertex per line (`VertexOutputFormat`) or one edge per line (`EdgeOutputFormat`). Following are some of the output formats that are available inside Giraph jars which you can use out of the box.

### 3.6.1   Vertex Output Formats

`VertexOutputFormat` is an abstract class which can be implemented to output the graph after computation. You can either write your own output format or can use the output formats that come packaged with Giraph. Following are some of the vertex output formats available in Giraph. Note that the `VertexOutputFormat` class for a Giraph job is identified by the value of property `giraph.vertexOutput FormatClass`.

#### 3.6.1.1   SimpleTextVertexOutputFormat

It is a text-based output format that is available in the Giraph's examples jar. It expects the vertex Id to be `LongWritable`, vertex value to be `IntWritable`, and edge value to be `FloatWritable`. Each line of the output text file contains a vertex id and its value-separated by the tab character.

### 3.6.1.2  `IdWithValueTextOutputFormat`

This output format is similar to the `SimpleTextVertexOutputFormat` but is more powerful. It does not impose any restriction on the vertex id or value type but uses their `toString()` methods to get their values. On each line of the output, it puts the id of a vertex and its value-separated by the tab character `'\t'` which is customizable. We can replace the default delimiter `'\t'` with any other character of our choice by specifying `output.delimiter` property either on the command line or `giraph-site.xml` file. If we want the value of the vertex to be written before its id on each line, we can do that by setting the `reverse.id.and.value` property to `true`.

### 3.6.1.3  `JsonBase64VertexOutputFormat`

This output format outputs the graph to a text file in a JSON format with base 64 encoded vertex ids, vertex values, and edge values. One line in the output file represents one vertex. Following is the output of our example shortest path project that we created in Sect. 2.9 using `JsonBase64VertexOutputFormat` instead of `IdWithValueTextOutputFormat`. All the ids and values look strange as they are base 64 encoded.

```
1  {"vertexId":"AAAAAAAAAAA=","edgeArray":["
      AAAAAAAAAAE/gAAA","AAAAAAAAAANAQAAA"],"
      vertexValue":"P/AAAAAAAA="}
2  {"vertexId":"AAAAAAAAAAE=","edgeArray":["
      AAAAAAAAAAA/gAAA","AAAAAAAAAJAAAAA","
      AAAAAAAAAAM/gAAA"],"vertexValue":"AAAAAAAAAA=
      "}
3  {"vertexId":"AAAAAAAAAAI=","edgeArray":["
      AAAAAAAAAFAAAAA","AAAAAAAAAARAgAAA"],"
      vertexValue":"QAAAAAAAAA="}
4  {"vertexId":"AAAAAAAAAAM=","edgeArray":["
      AAAAAAAAAABAQAAA","AAAAAAAAAAE/gAAA","
      AAAAAAAAAARAgAAA"],"vertexValue":"P/AAAAAAAA=
      "}
5  {"vertexId":"AAAAAAAAAAQ=","edgeArray":["
      AAAAAAAAAANAgAAA","AAAAAAAAAAJAgAAA"],"
      vertexValue":"QBQAAAAAAA="}
```

### 3.6.1.4  `JsonLongDoubleFloatDoubleVertexOutputFormat`

It is a VertexOutputFormat that supports JSON encoded vertexes featuring `Long Writable` ids, `DoubleWritable</code>` values and `FloatWritable` out-edge values/weights. Following is the output of our example shortest path project that we created in Sect. 2.9 using `JsonLongDoubleFloatDoubleVertexOutput Format` instead of `IdWithValueTextOutputFormat`.

```
1  [0,1,[[1,1],[3,3]]]
2  [1,0,[[0,1],[2,2],[3,1]]]
3  [2,2,[[1,2],[4,4]]]
4  [3,1,[[0,3],[1,1],[4,4]]]
5  [4,5,[[3,4],[2,4]]]
```

### 3.6.1.5 `GraphvizOutputFormat`

If you want to visualize the output graph of a Giraph job in GraphViz,[3] you can use this output format. It writes the graph to a dot file (graphviz format). A Giraph job can produce multiple output files depending on the number of workers. At the end of the job you can use the following to get a single graphviz file in HDFS:

```
hadoop fs -getmerge /hadoop/output/path data.txt
```

Following is the sample output for the shortest path algorithm in the graphviz dot file format:

```
0:id -> 1:id [label=1.0 ];
0:id -> 3:id [label=3.0 ];
"0" [label="<id> 0|1.0",shape=record,fillcolor="blue:orange"];
1:id -> 0:id [label=1.0 ];
1:id -> 2:id [label=2.0 ];
1:id -> 3:id [label=1.0 ];
"1" [label="<id> 1|0.0",shape=record,fillcolor="blue:orange"];
2:id -> 1:id [label=2.0 ];
2:id -> 4:id [label=4.0 ];
"2" [label="<id> 2|2.0",shape=record,fillcolor="blue:orange"];
3:id -> 0:id [label=3.0 ];
3:id -> 1:id [label=1.0 ];
3:id -> 4:id [label=4.0 ];
"3" [label="<id> 3|1.0",shape=record,fillcolor="blue:orange"];
4:id -> 3:id [label=4.0 ];
4:id -> 2:id [label=4.0 ];
"4" [label="<id> 4|5.0",shape=record,fillcolor="blue:orange"];
```

### 3.6.1.6 `AdjacencyListTextVertexOutputFormat`

This output format writes out the graph nodes as text, value-separated (by tabs, by default). With the default delimiter, a vertex is written written as:

```
<VertexId><tab><Vertex Value>[<tab><
    TargetVertexId><tab><EdgeValue>]+
```

[3]http://www.graphviz.org/.

The output graph of the shortest path example using `AdjacencyListText`
`VertexOutputFormat` can be written as:

```
0    1.0 1    1.0 3    3.0
1    0.0 0    1.0 2    2.0 3    1.0
2    2.0 1    2.0 4    4.0
3    1.0 0    3.0 1    1.0 4    4.0
4    5.0 3    4.0 2    4.0
```

### 3.6.1.7  `InMemoryVertexOutputFormat`

This output format is generally used for testing purposes. It stores the output in an
object of `TestGraph` class which you can access to verify the output or pass it on
to the next job using `InMemoryVertexInputFormat`. Before running the job,
it is `initializeOutputGraph(GiraphConfiguration conf)` method
should be called. When the job finishes, you can access the output graph using the
following code:

```
TestGraph outGraph = InMemoryVertexOutputFormat.
    getOutputGraph(); //get output graph
System.out.println(outGraph.toString()); //print
    output graph
```

## 3.6.2  Edge Output Formats

`EdgeOutputFormat` is an abstract class which can be implemented for writing a
custom output format. There are not many edge output formats available with Giraph
out of the box. We describe one of the mostly used edge output format in the following section. Programmers may utilize property `giraph.edgexOutputFormat`
`Class` to control the `EdgeOutputFormat` class.

### 3.6.2.1  `SrcIdDstIdEdgeValueTextOutputFormat`

This is an edge output format which writes out one edge per line. A line contains the source and destination id along with edge weight, but not the vertex
value. This is a demonstration output format to show the possibility of separately
outputting edges from vertexes. Default value separator is the `'\t'` character,
however it can be changed by specifying a different character for the property
`giraph.textoutputformat.separator` while running the job from command line or in `giraph-site.xml` file. The order of the values outputted on a
line can be reversed by setting the `giraph.textoutputformat.reverse`
property to `true`. Following is the output of the shortest path example using
`SrcIdDstIdEdgeValueTextOutputFormat`.

```
0    1    1.0
0    3    3.0
1    0    1.0
1    2    2.0
1    3    1.0
2    1    2.0
2    4    4.0
3    0    3.0
3    1    1.0
3    4    4.0
4    3    4.0
4    2    4.0
```

# Popular Graph Algorithms on Giraph

**4**

## 4.1 PageRank

PageRank [47] algorithm is a commonly used mechanism for identifying the significance or the authority of vertices in a graph. Therefore, it represents one of the most fundamental algorithms for Google Web Search. It is used to rank the internet web pages such that the important pages with higher Pagerank values are shown first in the search results. The basic idea is that the important websites likely receive more links than others. For example, CNN.com; a world leading news agency, is of high importance and there are many websites that have external links to CNN.

The PageRank value represents the likelihood that a person reaches a particular page by clicking randomly on links. The way the algorithm works is as follows: initially, the user is assumed to visit a certain web page; $P_1$, from a set of web pages. The probability of clicking a certain web page link is $I_1 = 1/$number of web pages. Once the first link is chosen, the next clicked link can be any web page that $P_1$ is referring to. If $P_1$ has $L_1$ links and one of the links is referring to $P_2$, then $P_1$ will pass $1/L_1$ of its importance to $P_2$. If $P_2$ has many links referring to it, then its importance $I_2$ will be the sum of the importance values of these incoming links to $P_2$.

The theory of PageRank [47] is based on the random surfer model, where a surfer always follow the outgoing links of the current web page. However, there is a chance that the surfer will stop clicking the current outgoing links and visits a new random page. To allow this in the PageRank calculation, there is a damping factor $p$, which is a decay factor. This $p$ value is usually specified to 0.85 which means that with a probability of 0.85, the surfer will follow an outgoing link of the current web page. The other 0.15 % of the time the surfer will randomly visit any other web page (with equal probability). This makes the PageRank equation as follows:

$$PR(p_i) = \frac{1-d}{N} + d \sum_{p_j \in M(p_i)} \frac{PR(p_j)}{L(p_j)} \tag{4.1}$$

© Springer International Publishing AG 2016
S. Sakr et al., *Large-Scale Graph Processing Using Apache Giraph*,
DOI 10.1007/978-3-319-47431-1_4

Notice that the PageRank values form a probability distribution over the set of web pages; i.e., the PageRank value is between 0 and 1 and the sum of the all web pages' ranks is one. Calculating PageRank values can be done using a simple iterative algorithm (see Sect. 4.1.2).

### 4.1.1   Example

Figure 4.1 shows a sample web graph where nodes represent web pages and edges represent their outgoing edges. The node *Bg* represents a personal blog which links to the owner Twitter *TW*, Facebook *FB*, and YouTube *YT* accounts. Similarly, a personal web page *WP* has outgoing links to *TW* and *FB* pages. To rank these different web pages, one could run PageRank algorithm and use the values as a way to measure the importance of different web pages.

Assume we want to run the PageRank algorithm on the graph shown in Fig. 4.1 for four supersteps.

- In the first superstep, there is an equal probability to visit any web page. Therefore, the initial value for each vertex is 0.2 (1/number of vertices). Then, each vertex sends a message to all its outgoing neighbors. The message value is decided based on the vertex value and the number of outgoing edges. For example, Page *WP* has a value of 0.2 and 2 outgoing edges; therefore, both pages *TW* and *FB* will be sent a message with value 0.1 (0.2/2) (see Fig. 4.2).
- In the second superstep, each node will sum the messages it receives. Then, it will apply the Eq. 4.1 to calculate the new vertex value. For example, vertex *TW* receives 0.1 and 0.066, so the sum is 0.166. The new vertex value is $(0.15/5) + 0.85 * 0.166 = 0.171$. Since *TW* has only one outgoing edge, it will send 0.171 (0.171/1) to its only outgoing neighbor; i.e., C.
- In the third superstep, vertices *WP* and *Bg* did not receive any message, so they became inactive.

**Fig. 4.1**  Sample web graph: Nodes are web pages while edges represent the outgoing links between them

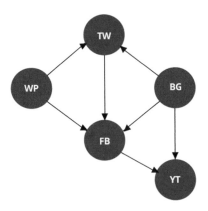

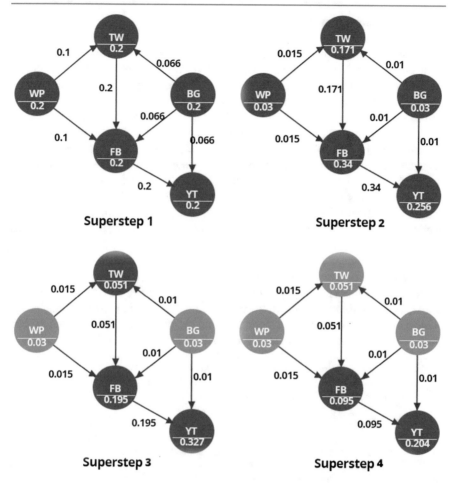

**Fig. 4.2** Running PageRank algorithm for four supersteps using the data graph shown in Fig. 4.1

- The process of receiving messages and applying the PageRank equation is repeated for the last two supersteps. At the end, each vertex maintains its PageRank value.

After running the PageRank algorithm, we can retrieve the PageRank values of all vertices. As expected the highest ranked node is *YB* as it has two incoming links one of them from another high ranked node; i.e., *FB*. Then, *FB* is the second high ranked vertex as it has 3 incoming links followed by *TW*, while the lowest ranked nodes are *WP* and *Bg* as they do not have any incoming links.

## 4.1.2  Implementation Details

We discuss next the implementation of the PageRank algorithm.[1] We first subclass the Vertex class to define a customized PageRankVertexComputation class. Recall that implementing any graph algorithm in Giraph requires extending the basic abstract class BasicComputation and specifying the four parameters for modeling the input graph: VertexID, VertexData, EdgeData, and MessageData. For PageRank algorithm, we specified these four parameters as follows: LongWritable, DoubleWritable, NullWritable, DoubleWritable for VertexID, VertexData, EdgeData, and MessageData, respectively (Line 1). Since we are not interested here with the edge data, it is specified as NullWritable. These four parameters should be changed according to the input graph and the algorithm specifications. Coding the PageRank algorithm is simple and straightforward. First, we have to specify the required number of supersteps (iterations) in advance (Line 2). In each superstep, we sum all the messages received from the direct neighbors of the current vertex (Lines 7–9). Then, we calculate the PageRank as shown in Eq. 4.1 (lines 10). Once the PageRank value is calculated, we update the current vertex value (Line 11). As long as the number of supersteps did not exceed the maximum number specified (Line 14), each vertex sends its updated value to all neighbors (lines 15–16). After five iterations (maximum number of supersteps defined here), all vertices will vote to halt and the program will terminate. At this stage, each vertex data represents its PageRank value.

```
1  public class PageRankVertexComputation extends
       BasicComputation<LongWritable, DoubleWritable,
       NullWritable, DoubleWritable> {
2      private static int MAX_SUPERSTEPS = 5;
3      @Override
4      public void compute(Vertex<LongWritable, DoubleWritable,
           NullWritable> vertex, Iterable<DoubleWritable>
           messages) throws IOException {
5          if (getSuperstep() >= 1) {
6              double sum = 0;
7              for (DoubleWritable message : messages) {
8                  sum += message.get();
9              }
10             DoubleWritable vertexValue = new
                   DoubleWritable((0.15f / getTotalNumVertices())
                   + 0.85f * sum);
11             vertex.setValue(vertexValue);
12         }
13
14         if (getSuperstep() < MAX_SUPERSTEPS) {
15             long edges = vertex.getNumEdges();
```

[1]The source code implementation is available on https://github.com/sakrsherif/GiraphBookSource Codes/blob/master/chapter04_05/src/main/java/bookExamples/ch4/algorithms/PageRankVertex Computation.java.

```
16          sendMessageToAllEdges(vertex, new
               DoubleWritable(vertex.getValue().get() /
               edges));
17      } else {
18          vertex.voteToHalt();
19      }
20   }
21 }
```

## 4.2 Connected Components

Two vertices are said to be connected if there is a path between them; i.e., they are reachable from each other. In an undirected graph, a connected component is a subgraph in which any two vertices are reachable from each other. A connected graph has exactly one connected component which is the graph itself. A vertex with no edges forms a connected component by itself. Finding the connected components in a graph has many applications. For example, finding connected components in a social network helps to identify the different communities of people who share the same interests. Other applications include clustering and graph indexing.

### 4.2.1 Example

Figure 4.3 shows a sample social graph which represents a set of users and their hobbies. It is obvious that this graph has three connected components. The first component contains the vertices Sarah, John, and Alice whose hobby is traveling while the second component contains the users interested in hiking; i.e., Tim and Joy. Notice that since the vertex Surfing is not connected to any other vertex, it formulates a connected component on its own.

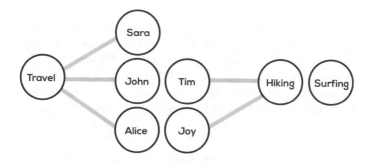

**Fig. 4.3** Sample social graph: Users associated with their interests

Running the connected components on the social graph in Fig. 4.3 requires four supersteps till the algorithm converges. Let us assume that each of the vertices in Fig. 4.3 is assigned a unique id. Figure 4.4 shows how the connected components algorithm is evaluated using the sample graph in Fig. 4.3. The algorithm works as follows:

- In the first superstep, each vertex initializes its connected component id (ccID) with its unique id. Then, each vertex sends its connected component id to all its neighbor vertices. For example, travel vertex sends its ccID (3) to vertices Sarah, John, and Alice. Similarly, it receives from them their ccIDs.
- In the second iteration, each vertex compares all the received ids with its ccID and updates the ccID to be the minimum value among all of them. For example, the vertex Sara received 3 and its ccID is 1, so its ccID remains 1. Since the ccID did not change, the vertex will not send to any neighbor vertex and will become inactive. Travel vertex, on the other hand, will change its ccID to the minimum between 3, 1, 2, and 4. It then will send its ccID (1) to its neighbors.
- In the third superstep, vertices John and Alice received a smaller ccID from travel vertex and update its ccID accordingly.
- In the fourth superstep, hiking vertex receives ccID of 5 which is not less than its ccID. At this step, hiking vertex becomes inactive. Since all vertices became inactive, the program will terminate. Notice that surfing vertex was active in the first superstep only and since it did not receive any message, it remains inactive throughout all the four supersteps.

After the program terminates, one can output the connected component id of all vertices and group similar ids together. We will find that vertices Travel, Sara, John, and Alice have the same connected component id of one and they form a connected component. Similarly, the second connected component (id = 5) contains the vertices Hiking, Tim, and Joy. The final connected component (id = 8) contains the vertex Surfing which is not connected to any other vertex (Fig. 4.4).

## 4.2.2  Implementation Details

We next describe the implementation of the connected components algorithm.[2] As shown below, we start by modeling the input graph by specifying the data types for the VertexID, VertexData, EdgeData, and MessageData. We use IntWritable to model VertexID, VertexData, and MessageData while NullWritable is used for EdgeData as we are not interested in edges' data in this algorithm (lines 1–2). Next, we write our compute function which takes as input a vertex and its incoming messages (Line 4).

---

[2]The source code implementation is available on https://github.com/sakrsherif/GiraphBookSource Codes/blob/master/chapter04_05/src/main/java/bookExamples/ch4/algorithms/Connected ComponentsVertex.java.

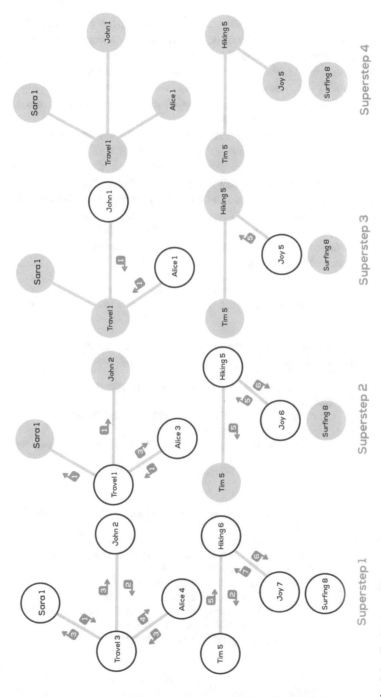

**Fig. 4.4** Evaluating connected components algorithm using the social graph shown in Fig. 4.3

Notice that the types specified here should match the types already mentioned when defining the class.

The first superstep is a special case. Each vertex initializes its connected component with its id (Line 4). Then it will iterate over all edges, compares its connected component id with the ids of its neighbors, and set the connected component id to the minimum id (lines 6–11). Now, each vertex maintains the minimum vertex id among itself and its neighbors. Then, the vertex broadcasts the connected component id to all neighbors; except those who have the same id as the minimum (lines 12–19). Then, the vertex votes to halt (Line 21).

In the remaining supersteps, each vertex compares its current component id with the received ids from its neighbors and selects the minimum value (lines 25–31). If the current component id of the vertex is smaller than all other ids, then there is no necessary change in the vertex state and the change flag is set to false. If the change flag is true, this means the connected component of the current vertex changed and the new id has to be propagated to all other neighbors (lines 32–35). This will continue for a number of supersteps until the algorithm converges; i.e., no more changes are found. At this stage, all vertices will be inactive and each vertex maintains its connected component id.

```
 1  public class ConnectedComponentsVertex extends
        BasicComputation<IntWritable, IntWritable, NullWritable,
        IntWritable>
 2  {
 3      public void compute(Vertex<IntWritable, IntWritable,
            NullWritable> vertex, Iterable<IntWritable> messages)
            throws IOException {
 4          int currentComponent = vertex.getValue().get();
 5          if (getSuperstep() == 0) {
 6              for (Edge<IntWritable, NullWritable> edge :
                    vertex.getEdges()) {
 7                  int neighbor = edge.getTargetVertexId().get();
 8                  if (neighbor < currentComponent) {
 9                      currentComponent = neighbor;
10                  }
11              }
12              if (currentComponent != vertex.getValue().get()) {
13                  vertex.setValue(new
                        IntWritable(currentComponent));
14                  for (Edge<IntWritable, NullWritable> edge :
                        vertex.getEdges()) {
15                      IntWritable neighbor =
                            edge.getTargetVertexId();
16                      if (neighbor.get() > currentComponent) {
17                          sendMessage(neighbor,
                                vertex.getValue());
18                      }
19                  }
20              }
21              vertex.voteToHalt();
```

```
22          return;
23       }
24       boolean changed = false;
25       for (IntWritable message : messages) {
26            int candidateComponent = message.get();
27            if (candidateComponent < currentComponent) {
28                 currentComponent = candidateComponent;
29                 changed = true;
30            }
31       }
32       if (changed) {
33            vertex.setValue(new
                  IntWritable(currentComponent));
34            sendMessageToAllEdges(vertex, vertex.getValue());
35       }
36       vertex.voteToHalt();
37    }
38 }
```

## 4.3  Shortest Path

Finding the shortest path is a well known problem in graph theory and network optimizations. The shortest path between a pair of vertices is the path that starts with the source vertex and ends at the target vertex such that the sum of weights of its constituent edges is minimized. We focus here on a slightly different problem; Single Source Shortest Path (SSSP) problem. SSSP finds the shortest path between a certain source vertex and all other vertices in the graph.

### 4.3.1  Example

Figure 4.5 shows a sample road network with five vertices representing different cities; Chicago, Boston, Philadelphia, Detroit and New York. Each edge has a direc-

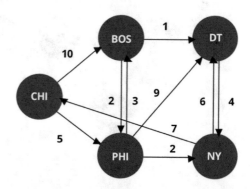

**Fig. 4.5** Sample road network: Nodes represent cities while edges represent the roads between them. Edge labels denote the cost to travel from one city to another

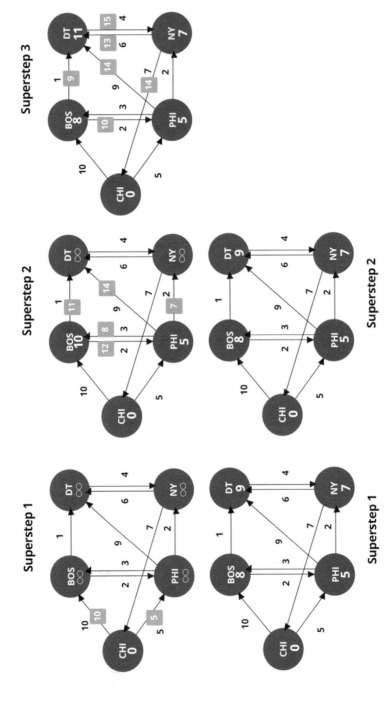

**Fig. 4.6** Evaluating Single Source Shortest Path (SSSP) algorithm using the sample road network in Fig. 4.5 starting from Chicago city

tion that defines its source and destination. Moreover, each edge has a weight which can be considered as the distance between the two cites. Assume that a user in Chicago wants to know what are all the shortest paths to the other four cities.

Initially, all vertices except the source assign its value to infinity which means that the vertex is initially unreachable. Then, each vertex finds the minimum distance to all its neighbors. Figure 4.6 shows how SSSP is evaluated using the sample road network in Fig. 4.5.

- In the first iteration, all vertices have minimum distance of infinity except the source vertex; CHI, which has a distance of zero to itself. If the found minimum distance is less than the vertex value, the vertex will propagate the minimum distance in addition to the edge weight. Only the source vertex in Fig. 4.5 is the one that satisfies this condition. Then, Chicago vertex will send 10 (0 + 10) and 5 (0 + 5) to vertices Boston and Philadelphia, respectively. Then, Chicago vertex votes to halt.
- In the second iteration, only vertices Philadelphia and Boston are active. For Philadelphia, it receives only one message (5) which is less than its value (Infinity). Therefore, it will update its value to the minimum distance (5) and propagate 8 (5 + 3), 14 (5 + 9), and 7 (5 + 2) to vertices Boston, Detroit, and New York, respectively. Similarly, Boston will propagate 12 and 11 to vertices Philadelphia and Detroit, respectively. Then both vertices Philadelphia and Boston will vote to halt.
- In the next superstep, vertices Boston, New York, and Detroit are active as they received messages. Each vertex will get the minimum distance between its value and the received messages and propagate the corresponding distances.
- This process continues till all vertices become inactive at the fifth superstep.

After all vertices vote to halt, each vertex maintains the distance required to travel from the source vertex; Chicago, to the present vertex. For example, the shortest path from Chicago to Boston has a cost equals 8 (vertex value of Boston). The path itself requires going from Chicago to Philadelphia (cost = 5) and from Philadelphia to Boston (cost = 3).

## 4.3.2   Implementation Details

We next describe how to implement the single source shortest path algorithm.[3] As usual, we start by defining the graph model (Line 1). Notice that, in this model we specified *FloatWritable* for modeling the EdgeData, which is supposed to maintain the distance between the two vertices at the edge sides. Next, we define the source

---

[3]The source code implementation is available on https://github.com/sakrsherif/GiraphBookSource Codes/blob/master/chapter04_05/src/main/java/bookExamples/ch4/algorithms/SimpleShortest PathsComputation.java.

vertex by using the LongConfOption. This will allow us to pass the source vertex id to the program via the command-line option *SimpleShortestPathsVertex.sourceId*. We next define a utility function which is responsible for deciding whether a vertex is the same as the source vertex or not (lines 3–5). To do so, it compares the id of the input vertex to the id specified in the *SOURCE_ID* parameter.

Initially, at the first superstep, each vertex initializes its own vertex value to the maximum possible value (*Double.MAX_VALUE*) to denote that it is initially unreachable (Line 11). Then, each vertex calculates the minimum cost of going to its neighbors (lines 12–14). If the vertex is the source vertex, then it has a minimum distance of zero as the source vertex is our starting point. Otherwise, the minimum distance is initialized to the maximum value. Then, we iterate over all received messages and find the minimum distance between the current vertex and any of its neighbors.

In the subsequent supersteps, each vertex compares its value to the minimum found distance considering the received messages (Line 15). If a smaller distance is found, the vertex propagates the smaller found distance to all its neighbors. Each vertex updates its value to the minimum found distance (Line 16). Then, for all its neighbors, it sends a value equal to the edge weight plus the smaller found distance (lines 17–20). This process continues for a number of supersteps till the program converges; i.e., each vertex already maintains the smallest possible distance. When all vertices are inactive, each vertex maintains the cost of going from the source vertex to the current vertex.

```
1  public class SimpleShortestPathsComputation extends
        BasicComputation<LongWritable, DoubleWritable,
        FloatWritable, DoubleWritable> {
2      public static final LongConfOption SOURCE_ID = new
            LongConfOption("SimpleShortestPathsVertex.sourceId",
            1);
3      private boolean isSource(Vertex<LongWritable, ?, ?>
            vertex) {
4          return vertex.getId().get() ==
                SOURCE_ID.get(getConf());
5      }
6      @Override
7      public void compute(Vertex<LongWritable, DoubleWritable,
            FloatWritable> vertex, Iterable<DoubleWritable>
            messages) throws IOException {
8          if (getSuperstep() == 0) {
9              vertex.setValue(new
                    DoubleWritable(Double.MAX_VALUE));
10         }
11         double minDist = isSource(vertex) ? 0d :
                Double.MAX_VALUE;
12         for (DoubleWritable message : messages) {
13             minDist = Math.min(minDist, message.get());
14         }
15         if (minDist < vertex.getValue().get()) {
16             vertex.setValue(new DoubleWritable(minDist));
```

```
17        for (Edge<LongWritable, FloatWritable> edge :
              vertex.getEdges()) {
18            double distance = minDist +
                  edge.getValue().get();
19            sendMessage(edge.getTargetVertexId(), new
                  DoubleWritable(distance));
20        }
21    }
22    vertex.voteToHalt();
23  }
24 }
```

## 4.4  Triangle Closing

In a social network graph, vertices represent users and edges represent connections (friendships) between users. Triangle closing is an algorithm that is responsible for finding for a certain user X, who are the other users that are not connected directly to X but they are most likely to be part his social circle. An obvious example for this in Facebook is "People you may know" side bar.

Triangle Closing algorithm works in only two supersteps. In the first superstep, each vertex $v$ sends its neighbor list to all of its neighbors. Then, the vertex $v$ votes to halt. In the second iteration, each vertex $v$ receives the neighbors list of its neighbors, so $v$ now maintains a two-hop adjacency information. Then, each vertex $v$ processes the 2-hop neighbor information and identifies the vertices who can form connections to the vertex. The list of new vertices are ranked in a descending order of the number of connections to neighbors of the vertex.

### 4.4.1  Example

Figure 4.7 shows a sample social graph where nodes represent users and edges represent friendship relations. One could use Triangle Closing to suggest for a user people who he might know.

Triangle Closing algorithm needs only two supersteps to produce the friends suggestions. We show in Fig. 4.8 how we evaluate the algorithm using the data graph in Fig. 4.7. The algorithm works in two supersteps:

- In the first iteration, each node send its entire neighbor list to all its neighbors. For example, *Alice* sends its neighbor list (i.e., *Bob, Roy*) to both neighbors Bob and Roy. As a result, in the next iteration, *Bob* knows the existence of *Roy* and vice versa. These three nodes all together form a triangle that we like to close.

**Fig. 4.7** Sample social graph: Nodes represent users while edges represent their friendship connections

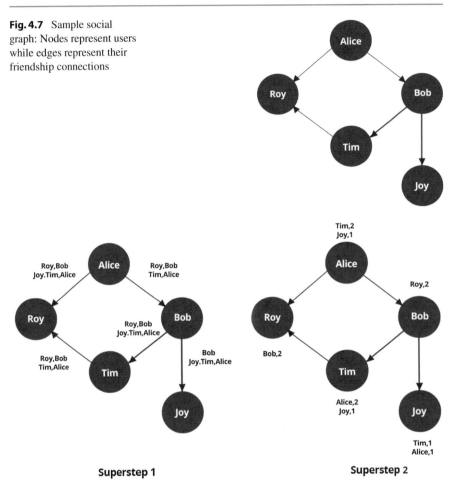

**Superstep 1**                                        **Superstep 2**

**Fig. 4.8** Evaluating Triangle Closing algorithm using the sample social network in Fig. 4.7

- In the second iteration, each node receives all neighbor ids as messages. For example, *Alice* received messages that contain the neighbors; Joy, Tim, and Alice. Assume that Alice will ignore the message that represent itself (i.e., Alice), then it received Tim two times and Joy once. This will form its ranked list of neighbors that Alice may know. Similarly, Bob will have Roy as a possible friend and vice versa. The algorithm will introduce Tim and Alice as new possible friends for Joy who currently is a friend of Bob only.

## 4.4.2   Implementation Details

We now show a simple implementation of Triangle closing algorithm.[4] The algorithm
assumes an undirected graph with no labels. Before we describe the code, we will
describe some utility classes. In this example, we define a custom vertex data class
which is a simple array list of int writables. The signature of the algorithm (Line 2)
states that the vertex value is defined as *IntArrayListWritable*; which will hold for
each vertex *v* the other vertices that *v* should connect to close the triangles with its
neighbors. The definition of *IntArrayListWritable* is shown in lines 35–62.

In the first superstep, each vertex sends all its neighbor list to all its neighbors
(lines 8–13). For each edge that the current vertex posses, send the neighbor id of
this edge to all other neighbors. Then, the vertex votes to halt (Line 34). In the
second superstep, each vertex processes its incoming messages. It will aggregate
these information in a hashmap *closeMap*, which will hold how many times the
vertex received each vertex id (Lines 14–17). In the rest of the algorithm, *closeMap*
will be sorted based on the frequency of each neighbor vertex. A new data structure
is defined (Line 20) to hold a sorted version of *closeMap*. We define *Pair* class in
lines 38–51. A pair has IntWritable key (vertex id) and integer value (count). It also
overrides the *Comparable.compareTo* method to facilitate sorting based on the value
(Lines 48–50). We simply iterate over *closeMap* and insert pairs of vertices with
their frequencies. The last step of the algorithm is to add the sorted vertices with
their frequencies to the vertex value (Lines 24–32). At the end of this algorithm,
each vertex holds the ranked set of vertices who are most likely to close triangles
with its neighbors.

```
1  package org.apache.giraph.examples;
2  public class SimpleTriangleClosingComputation extends
       BasicComputation<IntWritable,
       SimpleTriangleClosingComputation.IntArrayListWritable,
       NullWritable, IntWritable>
3  {
4    private Map<IntWritable, Integer> closeMap = Maps.<IntWritable,
         Integer>newHashMap();
5
6    @Override
7    public void compute(Iterable<IntWritable> messages) {
8      if (getSuperstep() == 0) {
9        // send list of this vertex's neighbors to all neighbors
10       for (Edge<IntWritable, NullWritable> edge : getEdges()) {
11         sendMessageToAllEdges(edge.getTargetVertexId());
12       }
13     } else {
14       for (IntWritable message : messages) {
```

---

[4]The source code implementation is available on https://github.com/sakrsherif/GiraphBookSource
Codes/blob/master/chapter04_05/src/main/java/bookExamples/ch4/algorithms/SimpleTriangle
ClosingComputation.java.

```
15      final int current = (closeMap.get(message) == null) ?0 :
            closeMap.get(message) + 1;
16      closeMap.put(message, current);
17    }
18    // make sure the result values are sorted and
19    // packaged in an IntArrayListWritable for output
20    Set<Pair> sortedResults = Sets.<Pair>newTreeSet();
21    for (Map.Entry<IntWritable, Integer> entry :
            closeMap.entrySet()) {
22      sortedResults.add(new Pair(entry.getKey(),
            entry.getValue())));
23    }
24    IntArrayListWritable outputList = new
            IntArrayListWritable();
25    for (Pair pair : sortedResults) {
26      if (pair.value > 0) {
27        outputList.add(pair.key);
28      } else {
29        break;
30      }
31    }
32    setValue(outputList);
33    }
34    voteToHalt();
35  }
36  public static class Pair implements Comparable<Pair> {
37    private final IntWritable key;
38    private final Integer value;
39    public Pair(IntWritable k, Integer v) {
40      key = k;
41      value = v;
42    }
43    public IntWritable getKey() { return key; }
44    public Integer getValue() { return value; }
45    @Override
46    public int compareTo(Pair other) {
47      return other.value - this.value;
48    }
49  }
50
51  public static class IntArrayListWritable extends
            ArrayListWritable<IntWritable> {
52    public IntArrayListWritable() {
53      super();
54    }
55    @Override
56    @SuppressWarnings("unchecked")
57    public void setClass() {
58      setClass(IntWritable.class);
59    }
60  }
61 }
```

## 4.5   Maximal Bipartite Graph Matching

A bipartite graph is a graph with two disjoint sets of vertices; left (L) and right (R) such that an edge connects a vertex from L to another vertex in R. For example, the left set can represent employees while the right set represents a possible set of tasks. A possible graph matching assigns employees to tasks. If we want to get as many assigned tasks as possible, then we are looking for a maximum matching.

In order to implement this algorithm in Giraph, we will provide a customized implementation for the VertexData and MessageData. The vertex value class will have a tuple of two values; a flag indicates whether the vertex has been matched and the id of the matching vertex on the other side. A set of access modifiers for setting and getting the values of the matching vertex are also supported.

```
public class VertexValue implements Writable {
  private boolean matched = false;
  private long matchedVertex = -1;
  public boolean isMatched() {
    return matched;
  }
  public long getMatchedVertex() {
    return matchedVertex;
  }
  public void setMatchedVertex(long matchedVertex) {
    this.matched = true;
    this.matchedVertex = matchedVertex;
  }
  . . .

}
```

Similarly, we define a custom MessageData class as follows. Each message maintains the id of the sender vertex and a message type. The type of the message can be one of the following: *MATCH_REQUEST* sent by left vertices, *REQUEST_GRANTED* to denote the grant of the match request between right and left vertices, and *REQUEST_DENIED* to denote the request denial.

```
public class Message implements Writable {
  private long senderVertex;
  private enum Type {
    MATCH_REQUEST,
    REQUEST_GRANTED,
    REQUEST_DENIED
  }

  private Message.Type type = Type.MATCH_REQUEST;
  public Message() {
  }
  public Message(Vertex<LongWritable, VertexValue,
      NullWritable> vertex) {
```

```
13    senderVertex = vertex.getId().get();
14    type = Type.MATCH_REQUEST;
15  }
16
17  public Message(Vertex<LongWritable, VertexValue,
        NullWritable> vertex,
18    boolean isGranting) {
19    this(vertex);
20    type = isGranting ? Type.REQUEST_GRANTED :
        Type.REQUEST_DENIED;
21  }
22  public long getSenderVertex() {
23    return senderVertex;
24  }
25  public boolean isGranting() {
26    return type.equals(Type.REQUEST_GRANTED);
27  }
28  . . . .
29 }
```

In this implementation, we assume that all vertices with even ids are in the left part while the odd ids are in the right. We start by modeling the input graph as follows. Notice that we used our customized classes; VertexValue and MessageData, to model the VertexValue and MessageData.

```
1 public class RandomizedMaximalMatchingComputation extends
2   BasicComputation<LongWritable, VertexValue, NullWritable,
        Message> {
```

Before we describe the algorithm, we first define a set of utility functions. *isLeft* determines whether the input vertex belongs to the left set of vertices or not. Similarly, *isRight* returns true if the input vertex belongs to the right set. *isNotMatchedYet* checks if the vertex is matched already. *createRequestMessage* creates a new matching request message for the input vertex. *createGrantingMessage* and *createDenyingMessage* create a grant and denial message to the matching request, respectively.

```
1
2   boolean isLeft(Vertex<LongWritable, VertexValue,
        NullWritable> vertex) {
3     return vertex.getId().get() % 2 == 1;
4   }
5   boolean isRight(Vertex<LongWritable, VertexValue,
        NullWritable> vertex) {
6     return !isLeft(vertex);
7   }
8
9   private boolean isNotMatchedYet(
10    Vertex<LongWritable, VertexValue, NullWritable> vertex) {
11    return !vertex.getValue().isMatched();
```

```
12   }
13
14   private Message createRequestMessage(
15     Vertex<LongWritable, VertexValue, NullWritable> vertex) {
16     return new Message(vertex);
17   }
18
19   private Message createGrantingMessage(
20     Vertex<LongWritable, VertexValue, NullWritable> vertex) {
21     return new Message(vertex, true);
22   }
23
24   private Message createDenyingMessage(
25     Vertex<LongWritable, VertexValue, NullWritable> vertex) {
26     return new Message(vertex, false);
27   }
```

The algorithm works in cycles; each cycle has four phases. To get the phase id, we find the superstep index mod 4. Then we check what is the phase id and act accordingly. In the first phase of a cycle, if the left vertex is not yet matched, it sends a message to all its neighbors requesting a match. After that, the vertex votes to halt. Notice that, this vertex will not be active again if it is already matched, does not have an outgoing edge or all its message recipients are already matched. If none of these conditions is true, the vertex will be active later.

```
1
2    public void compute(Vertex<LongWritable, VertexValue,
         NullWritable> vertex,
3    Iterable<Message> messages) throws IOException {
4      int phase = (int) (getSuperstep() % 4);
5      switch (phase) {
6        case 0:
7      if (isLeft(vertex)) {
8        if (isNotMatchedYet(vertex)) {
9          sendMessageToAllEdges(vertex,
             createRequestMessage(vertex));
10          vertex.voteToHalt();
11        }
12      }
13      // "If it sent no messages (because it is already matched,
             or has no
14      // outgoing edges), or if all the message recipients are
             already
15      // matched, it will never be reactivated. Otherwise, it
             will receive a
16      // response in two supersteps and reactivate."
17      break;
```

In the second phase, if the right vertex is not yet matched, the vertex will iterate over the incoming messages. The vertex randomly chooses a message and grants the request of the first one. Then, it sends a denial message to all others. After that, the vertex votes to halt.

```
 1      case 1: // "In phase 1 of a cycle,"
 2    // "each right vertex not yet matched"
 3    if (isRight(vertex)) {
 4      if (isNotMatchedYet(vertex)) {
 5        int i = 0;
 6        for (Message msg : messages) {
 7          Message reply;
 8          reply = (i == 0) ? createGrantingMessage(vertex) :
                createDenyingMessage(vertex);
 9          sendMessage(new LongWritable(msg.getSenderVertex()),
                reply);
10          ++i;
11        }
12        vertex.voteToHalt();
13      }
14    }
15    break;
```

In the third phase, a left vertex that is not yet matched will choose one of the grants it receives and sends an acceptance message. Then the vertex marks itself as matched and record its matching vertex.

```
 1    case 2:
 2      if (isLeft(vertex)) {
 3        if (isNotMatchedYet(vertex)) {
 4          for (Message msg : messages) {
 5            if (msg.isGranting()) {
 6              sendMessage(new LongWritable(msg.getSenderVertex()),
 7                createGrantingMessage(vertex));
 8              vertex.getValue().setMatchedVertex(msg.getSenderVertex());
 9              break;
10            }
11          }
12          vertex.voteToHalt();
13        }
14      }
15      break;
```

In the final phase of a cycle, a right vertex that is not yet matched will receive at most one acceptance message. Then it will mark itself as matched and record the matching vertex from the other side.

```
case 3:
    if (isRight(vertex)) {
      if (isNotMatchedYet(vertex)) {
        for (Message msg : messages) {
          vertex.getValue().setMatchedVertex(msg.getSenderVertex());
          break;
        }
      vertex.voteToHalt();
      }
    }
    break;
```

# Advanced Giraph Programming

**5**

## 5.1 Algorithm Optimization

In this section, we discuss alternative approaches to improve the flexibility of graph algorithms in Giraph and to improve their performance. The feasibility of these optimizations depends on the nature of the graph algorithm.

### 5.1.1 MasterCompute

The `MasterCompute` is an optional stage that performs centralized computation in Giraph. As shown in Fig. 5.1, `MasterCompute` class is executed in the master node at the beginning of each superstep and before starting the graph computation at the worker nodes. Programmers can use this stage to change graph computation classes, such as message combiners and vertex compute class, during runtime. Moreover, this stage gives developers access to the graph's information such as the superstep ID and the number of vertices and edges. `MasterCompute` can also be used to enable data sharing across different nodes. Users can use `GiraphConfiguration.setMasterComputeClass()` or property `giraph.masterComputeClass` to register their `MasterCompute` class. More details available in Giraph's documentation: https://giraph.apache.org/apidocs/org/apache/giraph/master/MasterCompute.html.

We show in Listing 5.1 an example implementation of `MasterCompute` class. The programmers have to implement four functions: `initialize()`, `compute()`, `readFields()`, and `write()`. `initialize()` is used to initialize objects or data sharing classes, such as aggregator, while `compute()` is the main function. `readFields()` and `write()` are used to enable Giraph to correctly serialize and deserialize the `MasterCompute` class. We also show in

© Springer International Publishing AG 2016
S. Sakr et al., *Large-Scale Graph Processing Using Apache Giraph*,
DOI 10.1007/978-3-319-47431-1_5

**Fig. 5.1** MasterCompute

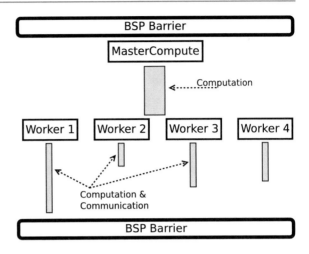

Listing 5.2 a `MasterCompute` class that dynamically changes the `compute` class and message combiner class between PageRank Sect. 4.1 and Connected Components Sect. 4.2 during runtime.[1]

```
Listing 5.1: Simple implementation of MasterCompute
1  public class simpleMasterCompute extends MasterCompute {
2      @Override
3      public void initialize() throws InstantiationException,
           IllegalAccessException {
4          // Initialization phase, used to initialize
               aggregator/Reduce/Broadcast
5          // or to initialize other objects
6      }
7      @Override
8      public void compute() {
9          //MasterCompute body
10         long ssID = getSuperstep();
11         long vCnt = getTotalNumVertices();
12         if(ssID > vCnt){
13             haltComputation();
14         }
15     }
16     public void readFields(DataInput arg0) throws IOException {
17         // To serialize this class fields (global variables)
               if any
18     }
19     public void write(DataOutput arg0) throws IOException {
```

---

[1]More examples of `MasterCompute` class can be found in https://github.com/sakrsherif/GiraphBookSourceCodes/tree/master/chapter04_05/src/main/java/bookExamples/ch5/masterCompute

```
20          // To deserialize this class fields (global variables)
               if any
21      }
22 }
```

**Listing 5.2:** MasterCompute with two compute classes

```
 1 public void compute() {
 2      // MasterCompute body
 3         long ssID = getSuperstep();
 4         if(ssID == 0){
 5             //At superstep 0, start with PageRank
 6             setComputation(PageRankVertexComputation.class);
 7             /*
 8              * If the number of edges 10 times larger than
 9              * the number of vertices, use a PageRank combiner
10              */
11             long vCnt = getTotalNumVertices();
12             long eCnt = getTotalNumEdges();
13             if(eCnt > 10 * vCnt){
14                 setMessageCombiner(sumDoubleCombiner.class);
15             }
16         }
17         else if (ssID == 10) {
18             //change the compute class in superstep 10
19             setMessageCombiner(minLongCombiner.class);
20             setComputation(ConnectedComponentsVertex.class);
21         }
22      }
```

## 5.1.2 Data Sharing Across Nodes

There are three types of data sharing across Giraph's nodes: broadcast, ReduceOperation, and Aggregators. For each data sharing approach has to be initialized in the MasterCompute using a unique String identifier.[2]

### 5.1.2.1 Broadcast

Broadcasting is the simplest way for the master node to communicate globally with worker nodes. It can be used to ensure that all vertices have access to the same information globally. A simple use case is when a user would like to assign random

---

[2]An example of an implementation of the Reduce and Aggregator functions can be found in https://github.com/sakrsherif/GiraphBookSourceCodes/tree/master/chapter04_05/src/main/java/bookExamples/ch5/dataSharing.

weights for the edges before starting PageRank, where the random generator depends on a custom generated seed. To ensure that all vertices use the same seed, programmers may produce the seed on the master `coumput` and then use `broadcast` to send the seed value to all vertices.

Broadcast is initialized in `MasterCompute` using the function call `broadcast (String, org.apache.hadoop.io.Writable)`. Once a broadcast object is initialized at the `MasterCompute` stage, Giraph sends one copy of this object to each worker. Only the master node is allowed to modify the broadcast object, while workers have read-only access to it through the function call `getBroadcast(String)`. Note that once the object is initialized it is broadcasted to all workers and, unless modified, Giraph will not broadcast the object again. Programmers can use Hadoop's datatypes, such as LongWritable, or use their custom datatype by implementing Hadoop's `Writable` interface. Please refer to the following link for more details: https://hadoop.apache.org/docs/stable/api/org/apache/hadoop/io/Writable.html. We show in Listing 5.3 an example of how to implement Hadoop's `Writable`, where we show an implementation for a `TextArrayWritable`.

Listing 5.3: Custom ArrayWritable class

```
public static class TextArrayWritable extends ArrayWritable {
    public TextArrayWritable() {
        super(Text.class);
    }

    public TextArrayWritable(String[] strings) {
        super(Text.class);
        Text[] texts = new Text[strings.length];
        for (int i = 0; i < strings.length; i++) {
            texts[i] = new Text(strings[i]);
        }
        set(texts);
    }
}
```

### 5.1.2.2 ReduceOperation

ReduceOperation is used to allow workers to communicate with the master node. Consider the random seed example in Sect. 5.1.2.1, where the seed generation may depend on the min and max vertex out-degree values. To send the min and max out-degree information to the master, each vertex would use ReduceOperation to send its out-degree count. In this case, the ReduceOperation picks the min and max values from the different out-degree values and reports it to the master. The master `compute` then generates the random seed from the incoming out-degree information, which is shared to the vertices through the broadcast operator.

ReduceOperation has to be initialized at the `MasterCompute` node first by using the function call `registerReduce(String, ReduceOperation<S,R>)`

or `registerReduce(String, ReduceOperation<S,R>, R)`, where $S$ is any Java Object and $R$ is an object that extends Hadoop's `Writable` interface. The implementation of ReduceOperation has to be commutative and associative to allow different workers to use the ReduceOperation in parallel without a specific order. Worker nodes can only use the ReduceOperation to update its value through function call `reduce(String, Object)`, where the master node is the only node that has a read access to the ReduceOperation. In other words, Giraph does not broadcast the ReduceOperation's class to all worker unless needed. Note that programmers have to make sure that their reduce function is thread safe since a worker can execute the same reduce class in parallel. ReduceOperation requires programmers to implement an initialization method `createInitialValue()`, `reducePartial()` to return partial reduced value, and `reduceSingle()` which is used by the workers. More details for the Reduce function can be found at the following link: https://giraph.apache.org/apidocs/org/apache/giraph/reducers/ReduceOperation.html. We show in Listing 5.4 an example of a sum `reduce` class.

Listing 5.4: Example of a reduce class

```
1  public class SUMReduce implements ReduceOperation<Long,
        LongWritable> {
2
3      public void readFields(DataInput arg0) throws IOException {
4          // To deserialize this class fields (global variables)
                if any
5      }
6
7      public void write(DataOutput arg0) throws IOException {
8          // To serialize this class fields (global variables)
                if any
9      }
10
11     public LongWritable createInitialValue() {
12         return new LongWritable(0);
13     }
14
15     public LongWritable reduceSingle(LongWritable curValue,
            Long valueToReduce) {
16         return new LongWritable(curValue.get() +
                valueToReduce);
17     }
18
19     public LongWritable reducePartial(LongWritable curValue,
            LongWritable valueToReduce) {
20         return new LongWritable(curValue.get() +
                valueToReduce.get());
21     }
22 }
```

### 5.1.2.3 Aggregators

Aggregators are functions that can be used to communicate across all nodes in Giraph. Both master and worker nodes can access and modify the value of an aggregator function. Again, consider the random seed example in Sects. 5.1.2.1 and 5.1.2.2; but assumes that PageRank also requires access to the min and max out-degree values. Users may use two different broadcast values, one for the random seed and another for the min and max out-degree information, or use aggregators to allow access to the min and max values for both the master `compute` and the vertex `compute` classes.

Similar to ReduceOperation, aggregator function has to be initialized by the master node, and it has to be commutative and associative. Functions registerAggregator (String, Class<? extends Aggregator<A>>) and registerPersistentAggregator(String, Class<? extends Aggregator<A>>) are used by `MasterCompute` to register an aggregator, while getAggregatedValue(String) and setAggregatedValue(String name, A value) are used to access and modify the aggregator value at `MasterCompute`. Once an aggregator is initialized, Giraph broadcasts a copy of the aggregator's object to each worker. Workers will have read-only access to the current aggregator value that has been set by the master. Workers can apply a value change to the aggregator object, by using the function call aggregate(String, A), where all changes will be synchronized and applied on the aggregator object by the end of the superstep. Both the `MasterCompute` of the next superstep will have a consistent view of the latest change in the aggregator object that was modified by the workers of the last superstep. If the `MasterCompute` of the new superstep did not alter the value of the aggregator, the workers would also be able to access the latest change in the aggregator object that was updated by the last superstep. There are two types of aggregators: nonpersistent and persistent aggregators. The nonpersistent aggregators will be reset to their default initial value at the end of the superstep before applying any value changes from workers of the current superstep. The persistent aggregators, on the other hand, will not reset the aggregator value and it will apply any changes requested by the workers of the current superstep on the latest aggregator value. Since the aggregator value is made available to both master and worker nodes, Giraph will keep synchronizing its value across workers even if no worker in the next superstep requires to access the aggregator value. We show in Fig. 5.2 an example of the workflow of a nonpersistent aggregator function. We show in Listing 5.5, an example of how to implement an aggregator class. To implement an aggregator, programmers have to implement five functions: aggregate(A), createInitialValue(), getAggregatedValue(), reset(), and setAggregatedValue(A). https://giraph.apache. org/apidocs/org/apache/giraph/aggregators/Aggregator.html.

Compared to other data sharing approaches, users can write aggregators values to disk. To allow this feature, users have to define the aggregator writer class through property `giraph.aggregatorWriterClass`. The default aggregator writer class is `TextAggregatorWriter`, however, users can implement their custom writers. When using this writer, users have to define the writing frequency and filename for the writer. The frequency of writing aggregators values to disk is controlled through property `giraph.textAggregatorWriter.frequency`, where it

**Fig. 5.2** Simple aggregator

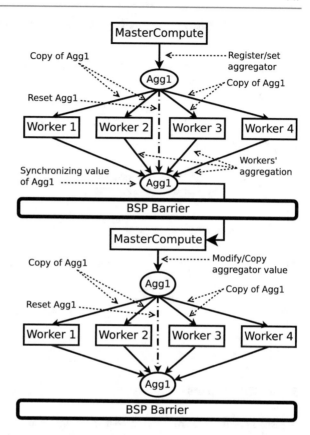

can be assigned one of the following values: An integer value, ALWAYS, AT_THE_END, or NEVER. An integer value defines after how many supersteps the writer should write the aggregator values to disk. For example, if a user assigned value 3 for property giraph.textAggregatorWriter.frequency, Giraph will write the values of the aggregators after every 3rd superstep. On the other hand, ALWAYS will make the aggregator to write the values to disk after every superstep while AT_THE_END delays the writing after completing all supersteps. Using NEVER, however, stops the aggregator writer from writing anything. Finally, users may customize the file name for the aggregator writer through property giraph.textAggregatorWriter.filename.

Listing 5.5: Example of a aggregator class

```
1  public class SUMAggregator implements Aggregator<LongWritable>
      {
2
3      LongWritable myValue;
4
5      public void aggregate(LongWritable value) {
```

```
6         myValue.set(myValue.get() + value.get());
7     }
8
9     public LongWritable createInitialValue() {
10        return new LongWritable(0);
11    }
12
13    public LongWritable getAggregatedValue() {
14        return myValue;
15    }
16
17    public void setAggregatedValue(LongWritable value) {
18        myValue.set(value.get());
19    }
20
21    public void reset() {
22        if (myValue == null) {
23            myValue = new LongWritable(0);
24        } else {
25            myValue.set(0);
26        }
27    }
28 }
```

### 5.1.3  Combiners

Message combiners are applied on the incoming messages for each vertex before executing the compute functions of the next superstep. The job of the combiner is to reduce the size of the incoming messages to a vertex by combining their values using a user-defined function. The function iterates over a list of incoming messages and produces one or more messages. It is preferred that the combiner function is commutative and associative to ensure correctness of the incoming messages to the vertex. Message combiners are similar to Reduce Sect. 5.1.2.2 Aggregate Sect. 5.1.2.3 functions, where users receive the vertex ID, original message and a new message and also required to define an initial, neutral, message. To combine, users have to modify the content of the original message concerning the new message. Combiners can be used to optimize the performance of PageRank Sect. 4.1, connected components Sect. 4.2 and shortest path Sect. 4.3. Notice that these three algorithms iterate over the messages and compute the sum of the incoming messages (in PageRank) or the minimum value of the incoming messages (in connected components and shortest path). As a result, we can easily move the computation of the sum or the minimum to the combiner function instead of the `compute` function. On the other hand, triangle closing Sect. 4.4 and maximal bipartite graph matching Sect. 4.5 cannot use combiners. This is because triangle closing requires access to all messages to sort them, while maximal bipartite graph matching has distinct actions for the incoming messages.

We show in Listing 5.6 an example of a combiner class that finds the minimum value (line 7), which may be used for connected components and shortest path. Note that the classes for Vertex ID and the messages must be identical to vertex ID and message classes of user's `Computation` class. To add a message combiner, users have to register their combiner class using method `GiraphConfiguration.setMessageCombinerClass()` or through defining the path to their combiner class in property `giraph.messageCombiner Class`. Users can also use Giraph's default message combiners: `DoubleSum MessageCombiner`, `FloatSumMessageCombiner`, `MinimumDouble MessageCombiner`, `MinimumIntMessageCombiner`, `SimpleSum MessageCombiner`. Note that at the current implementation of Giraph, you can only use a single combiner class for your Giraph's job.

Listing 5.6: Example of a combiner class

```
public class minLongCombiner
    implements MessageCombiner<LongWritable, LongWritable> {
  @Override
  public void combine(LongWritable vertexIndex,
          LongWritable originalMessage, LongWritable
              messageToCombine) {
    if (originalMessage.get() > messageToCombine.get()) {
      originalMessage.set(messageToCombine.get());
    }
  }

  @Override
  public LongWritable createInitialMessage() {
    return new LongWritable(Long.MAX_VALUE);
  }
}
```

## 5.1.4  Coarse-Grained Processing

Pregel is originally designed for fine-grained graph processing, where graph vertices run the user's graph algorithm independently. On top of vertex-based computation, users can also use coarse-grained on each worker which is known as worker context. The worker context is very similar to MasterCompute where all workers have access to globally defined aggregators, reducers, and broadcasted values. Moreover, each worker can execute a set of user-defined functions before and after each superstep and exchange messages with other workers. Users can extend Giraph's default worker context class `DefaultWorkerContext` or define their own, where the worker context is controlled through property `giraph.workerContextClass`.

A worker context has four main functions: `preApplication()`, `preSuperstep()`, `postApplication()` and `postSuperstep()`. `preApplication()` is considered as an initialization step that is executed before starting the first superstep in Giraph, while `postApplication()` is executed after the last superstep. During execution, the function `preSuperstep()` is executed before starting the next superstep while `postSuperstep()` is executed after finishing the current superstep. Each worker has access to the data sharing functions (`aggregate()`, `getAggregatedValue()`, `getBroadcast()` and `reduce()`) and computation information (`getSuperstep`, `getTotalNum Vertices` and `getTotalNumEdges`). Users can use functions `getMyWorker Index()` and `getWorkerCount()` to know the ID of the current worker and the total number of available workers. The worker ID is used to send messages to specific workers by using function `sendMessageToWorker()`. Moreover, function `getAndClearMessagesFromOtherWorkers()` is used to read all messages from other workers.

A direct use case for the worker context is to gain some knowledge on the graph after the user's graph algorithm. For example, assume that we would like to know if the input graph is disconnected or not after executing Connected Components. To achieve this, we use an aggregator and a reduce operator. The aggregator reports the global minimum components ID in the first superstep, while the reduce operator allows vertices to report whether the aggregator value matches their final minimum components ID or not. The reduce operator, in this case, is a Boolean AND operator. After that, we read the reduce operator value in function `postApplication()`; the graph is concluded to be connected if the value is `true`, otherwise the graph is disconnected.

## 5.2   Dealing with More Complex Graph Algorithms

Not all graph algorithms are naturally implementable in Giraph because its API and its traditional optimizations (like in Sect. 5.1) may not be applicable. Some algorithms require undirected graph computations, while others require fine-grained synchronization between adjacent vertices. Since Giraph's computing model is directed and distributed; it does not naturally support undirected graphs or have the ability to synchronize value assignment to adjacent vertices. To solve this problem, researchers tend to modify Pregel API to serialize the value assignment process through the master  compute, as suggested by the work in [48, 49]. In this section, we show how to handle undirected graphs and discuss nontraditional fine-grained vertex synchronization without the need to modify Giraph API.

### 5.2.1   Graph Mutations

During runtime, programmers can change the structure of the graph globally by adding or removing vertices and edges. Each vertex in Giraph, as described in Sect. 3.1.3, through `Computation` interface, can add new vertices and edges using `addVertexRequest()` and `addEdgeRequest()`, while methods `removeVertexRequest()` and `removeEdgesRequest()` are used to remove vertices and edges. All add and remove requests are applied on the graph just before starting the next superstep. The behavior of vertex additions and deletions is controlled by `DefaultVertexResolver` class while edge modifications are controlled by `OutEdges` class through `DefaultVertex` class. A new vertex is created if an `addVertexRequest()` is received. Moreover, if a nonexisting vertex receives a message, Giraph will automatically add the nonexisting vertex to the graph unless the user assigns the value of property `giraph.vertex.resolver.create.on.msgs` to false. Moreover, if a new vertex add with edges request is received, Giraph builds the new vertex with the outgoing edges and pass it to `DefaultVertexResolver` class to be added. However, if `addVertexRequest()` is issued for an existing vertex, the `DefaultVertexResolver` ignores that request even if the new vertex has edges that do not exist in the original vertex. Remove edge requests are also ignored if the source vertex is a nonexisting vertex. If the source vertex exists, however, remove edge requests are passed to the defined `giraph.inputOutEdgesClass` (or `giraph.utEdgesClass`) of that vertex. The default behavior of the `OutEdges` class is to add an edge even if it exists and removes an edge if it exists. However, users can define their own `OutEdges` class and control how edges are manipulated (Sect. 5.3.3). The `DefaultVertexResolver` class uses a resolver mechanism (implemented by method `resolve()`) to ensure the graph's consistency by controlling the order of processing conflicting removal and adding requests for the same vertex. The default implementation of `resolve()` processes edge removals first then vertex removals. After completing all removal requests, it then processes vertex addition requests, and after that, it completes edge additions requests. For communication optimization purposes, Giraph uses the configuration option `giraph.maxMutationsPerRequest` to control how many mutation requests the graph partition should wait for before sending those requests to other partitions. The default value of this configuration is 100; programmers should choose the best value for this configuration based on their application because a smaller value may increase the communication overhead while a bigger value may reduce the response time of graph mutations. Note that users can also use `addEdge()` and `removeEdges()` using the `Vertex` interface, as described in Sect. 3.1.2, however, those changes only affect the out-degree edges of the vertex.

## 5.2.2   Undirected Graphs

By default, a vertex in Giraph only stores the IDs of its outgoing edges. Some graph algorithms, such as connected components (Sect. 4.2), require undirected graphs to correctly analyze it. Consider the following vertex file:

```
2    1    3
```

From the above file, Giraph will generate three vertices, vertex 1, 2, and 3 as shown in Fig. 5.3, where vertex 2 stores 1 and 3 as edges while vertices 1 and 3 do not store any edge. If this graph was intended to be undirected, such as the graph in Fig. 5.4, running the connected components on top of Fig. 5.3, assuming it represents the undirected graph in Fig. 5.4, will return wrong analysis. That is, the connected components algorithm on Fig. 5.3 result in three connected components while on Fig. 5.5 it results in one connected component. To avoid such limitations, users are required to represent each undirected edge as two directed edges in the input file as follows:

```
2    1    3
1    2
3    2
```

With the above vertex file, Giraph will generate three vertices, vertex 1, 2, and 3, where vertices 1 and 3 store 2 as an edge, and vertex 2 stores 1 and 3 as edges (Fig. 5.5). Instead of manually editing the input file, users can automate the process of generating the required extra edge by either using class SccLongLongNullTextInput Format (Sect. 3.5.4.5) as a vertex reader or class ReverseEdgeDuplicator as an edge reader. To generate the extra edges with SccLongLongNullTextInput

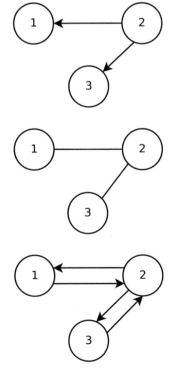

**Fig. 5.3**  A directed graph with two outgoing edges from vertex 2

**Fig. 5.4**  An undirected graph with two edges

**Fig. 5.5**  A directed graph that best describes within Giraph the graph in Fig. 5.4

Format, the input file should be in an adjacency list format. After Giraph loads the graph into memory, users have to add the extra edge manually, through addEdge Request() (Sect. 5.2.1), at each vertex by using the in-degree edge information stored in class SccVertexValue. We show in Listings 5.7 and 5.8 code snippets on converting the directed graph into an undirected graph through addEdgeRequest() and addEdge(). Since the graph is already established in memory, users can avoid adding duplicate edges by scanning the current edges in each vertex before requesting Giraph to add a new edge. The disadvantage of this class is that the graph algorithm has to start on the second superstep since the first one is dedicated to establishing new edges. Note that there is no difference between Listings 5.7 and 5.8 because addEdgeRequest() is applied at the next superstep. On the other hand, to use ReverseEdgeDuplicator the input file has to be in edge list format. When using this class as an edge reader, each incoming edge from the input file is automatically duplicated in reverse. Unlike SccLongLongNullTextInputFormat, this class does not waste a superstep since it automatically adds new edges during graph loading and before starting the first superstep. The disadvantage of this class is, however, that duplicate edges may exist because this class does not have the full picture of the graph before creating duplicate reverse edges.

**Listing 5.7:** Converting a directed graph into undirected graph by adding new edges with compute.addEdgeRequest()

```
1  public void compute(Vertex<LongWritable, LongWritable,
       LongWritable>
2      vertex, Iterable<LongWritable> messages) throws
           IOException {
3      // add remote edges
4      LongWritable myVertexID = vertex.getId();
5      if (ss == 0) {
6          for(Edge<LongWritable,Text> edge: vertex.getEdges()){
7              super.addEdgeRequest(edge.getTargetVertexId(),
8                  EdgeFactory.create(myVertexID, new Text()));
9          }
10     }
11     ....
12 }
```

**Listing 5.8:** Converting a directed graph into undirected graph by adding new edges with vertex.addEdge()

```
1  public void compute(Vertex<LongWritable, LongWritable,
       LongWritable>
2      vertex, Iterable<LongWritable> messages) throws
           IOException {
3      // send to all nbrs my ID
4      LongWritable myVertexID = vertex.getId();
5      if (ss == 0) {
```

```
 6          super.sendMessageToAllEdges(vertex, outMessage);
 7      }
 8      // Get nbr IDs and add local edges
 9      else if (ss == 1) {
10          Iterator<LongWritable> msgIterator =
                messages.iterator();
11          while (msgIterator.hasNext()) {
12              LongWritable removeVertexID = new
                    LongWritable(msgIterator.next().get());
13              vertex.addEdge(EdgeFactory.create(removeVertexID,
                    new Text()));
14          }
15      }
16      ....
17 }
```

### 5.2.3  Synchronizing the States of Neighboring Vertices

Some graph algorithms, such as graph coloring and vertex cover, require that adjacent vertices must have different value assignment. It is easy to achieve this in a centralized setting, but in Giraph it is not possible to serialize the value assignment. For example, let us assume that we want to assign some random values to the vertices of the graph of Fig. 5.6 such that no two adjacent edges have the same random value. In Fig. 5.7 we show the random value assignment, for the vertices in Fig. 5.6, where vertex **B** conflicts with vertex **E**. To solve this problem, users can imitate collision detection and avoidance techniques from networking field in their compute() function. The collision detection and avoidance algorithm work in three stages: value assignment, collision detection, and collision resolve.

Figure 5.7 shows the value assignment stage at superstep 1. In this superstep, all vertices send their random number assignment to their neighbors. In superstep 2 (Fig. 5.8), vertex **E** detects that its value conflicts with the value of vertex **B**, however, vertex **B** does not know about the conflict. As a result, only vertex **E** starts a back-off process in the next superstep where it removes its current value assignment and stays idle for some supersteps. The back-off period can be based on random or Exponential [50] value assignment. It is important that different vertices

**Fig. 5.6**  Example figure

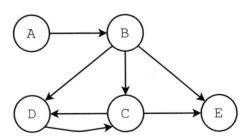

**Fig. 5.7** Superstep 1: Value assignment

**Fig. 5.8** Superstep 2: Collision detection between vertices **B** and **E**

**Fig. 5.9** Supersteps 3–5: Collision resolve; vertex **E** in back-off period

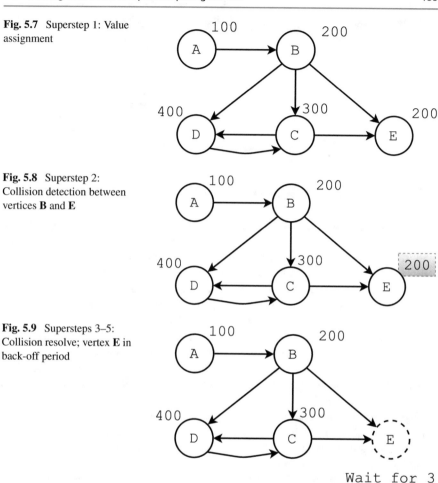

Wait for 3 supersteps

have different back-off values to minimize recurrent collisions. On the other hand, all other vertices, i.e., **A**, **B**, **C** and **D**, finalize their value assignment because they did not receive any conflicting messages as shown in Fig. 5.8. Vertex **E**, however, stores that 200 and 300 have been used and cannot be used as a choice later. In Fig. 5.9 we show that vertex **E** becomes idle for 3 supersteps. Note that during the back-off period, vertex **E** still receives value assignments from other neighbors (if any) where it must store used values by other vertices. Finally, we show in Fig. 5.10 that vertex **E** assigns a new random value without conflicting with others. The value assignment completes when all vertices halt and do not send any messages. Note that this approach is a random computing model which does not necessarily guarantee optimality.

**Fig. 5.10** Supersteps 6:
New assignment for vertex **E**

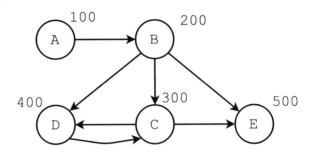

## 5.2.4   Implementing Graph Coloring

Graph coloring is a common graph problem that assigns distinct labels (colors) to adjacent vertices (or edges). It is an important algorithm that has been used in many real life applications for scheduling different events. One of the most common use cases for this algorithm is to schedule university exams to ensure that no two exams having common students are scheduled at the same time. Graph coloring can also be used in social networks to improve friend matching and cliques discovery. Although graph coloring is very easy to be implemented in serial settings, graph coloring is very tricky to be applied in Giraph. Due to Giraph's distributed nature, there is no guarantee on the order of executing the vertices `compute` function which makes it difficult to synchronize the color assignment between different vertices. Moreover, graph coloring is mostly applied on undirected graphs, while Giraph only supports directed graph processing.

To implement graph coloring In Giraph, we utilize two optimizations: (1) converting the directed input graph into undirected graph Sect. 5.2.2, and (2) synchronizing the color assignment between adjacent vertices using distributed collision detection technique Sect. 5.2.3.[3] We show the code of graph coloring algorithm in Listing 5.9. Assume that we want to apply the graph coloring algorithm on the graph in Fig. 5.11, where this graph represents university computer science courses with common students. The first step is to convert the input from a directed graph into an undirected graph. In Superstep 0, all vertices send their ID to their neighbors defined by their out-degree edges (Lines 7–10). Then, in Superstep 1 each vertex gets its neighbors ID and then submits a local `addEdge()` to generate the undirected graph (Lines 12–18) as shown in Fig. 5.12. In the same superstep, the graph coloring algorithm starts. We assume that we are using an ordered set of the following colors: Green, Blue, and Red. In other words, all vertices in Superstep 1 will try to assign the color Green (denoted by 0 in Line 19) and send the selected color to its neighbors (Line 20) as shown in Fig. 5.13. In Superstep 2, all vertices will discover that the chosen color

---

[3]The code is available in: https://github.com/sakrsherif/GiraphBookSourceCodes/blob/master/chapter04_05/src/main/java/bookExamples/ch5/algorithms/GraphColoring.java.

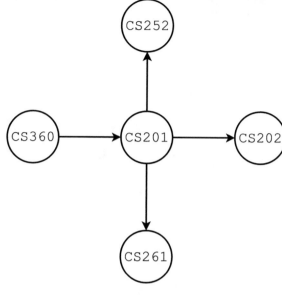

**Fig. 5.11** Example directed graph of courses with common students

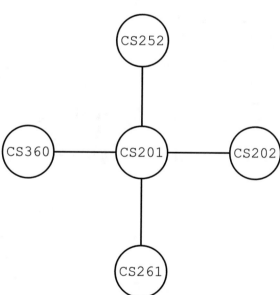

**Fig. 5.12** Example undirected graph of courses with common students

**Fig. 5.13** Superstep 1:
Color assignment

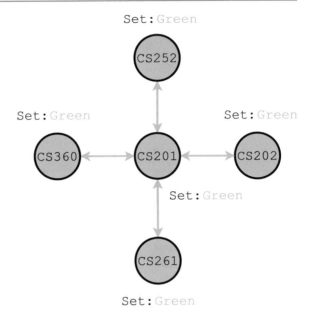

**Fig. 5.14** Superstep 2:
Collision detection; all
vertices revert to random
back-off state

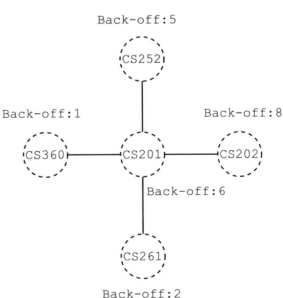

cannot be used because it conflicts with their neighbors (Lines 38–40). As a result, all vertices starts the random back-off state as shown in Fig. 5.14 and decrement 1 from the back-off count in each next superstep (Lines 48–50). In Superstep 3, vertex CS360 gets out of the back-off state, selects the color Green again, and sends its choice to its neighbors as shown in Fig. 5.15. In Superstep 4, vertex CS360 confirms the color green while CS201 gets out of the back-off state (line 30) because it received a color update. In this case, vertex CS201 picks the next available color, which is Blue because it knows that Green have been used previously (Line 56). At the same time, vertex CS261 happens to complete the back-off period and selects the color Green. Both vertices CS201 and CS261 send their choice to their neigh-

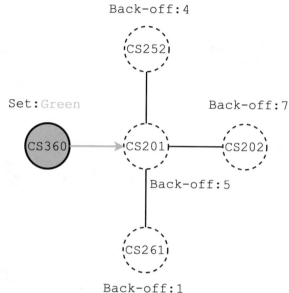

**Fig. 5.15** Superstep 3: Vertex CS360 becomes active

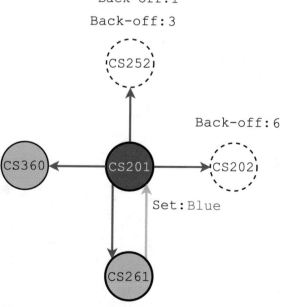

**Fig. 5.16** Superstep 4: Vertices CS201 and CS261 become active

bors as shown in Fig. 5.16. Similarly, in Superstep 5, vertices CS252 and CS202 become active due to the color assignment received from vertex CS201, select the color Green and send to their neighbors their choice as shown in Fig. 5.17. Finally as shown in Fig. 5.18, by Superstep 6 all vertices have finalized their color choice and the algorithm terminates.

**Fig. 5.17** Superstep 5:
Vertices CS202 and CS252
become active

**Fig. 5.18** Superstep 6: All
vertices are done with color
assignment

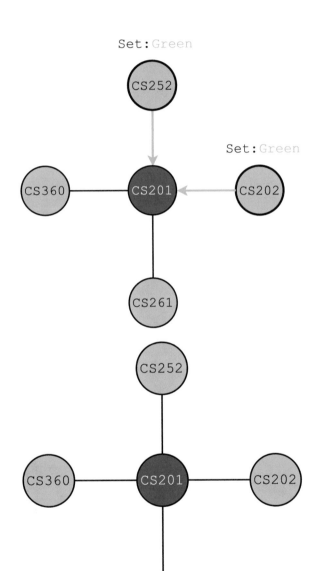

## Listing 5.9: Graph coloring algorithm

```
1  public void compute(Vertex<LongWritable, Text, Text> vertex,
2          Iterable<Text> messages) throws IOException {
3      long ss = super.getSuperstep();
4
5      // send to all nbrs my ID
6      LongWritable myVertexID = vertex.getId();
7      if (ss == 0) {
8          Text outMessage = new Text(myVertexID.get() + "");
9          super.sendMessageToAllEdges(vertex, outMessage);
10     }
11     // get nbr IDs
12     else if (ss == 1) {
13         Iterator<Text> msgIterator = messages.iterator();
14         while (msgIterator.hasNext()) {
15             LongWritable removeVertexID = new LongWritable(
16                 Long.parseLong(msgIterator.next().toString()));
17             vertex.addEdge(EdgeFactory.create(removeVertexID, new
                   Text()));
18         }
19         myColor = 0;
20         super.sendMessageToAllEdges(vertex, new Text(myColor +
               ""));
21     } else {
22         Iterator<Text> msgIterator = messages.iterator();
23         // check incoming messages
24         if (IamDone) {
25             while (msgIterator.hasNext()) {
26                 msgIterator.next();
27             }
28         } else if (msgIterator.hasNext()) {
29             backOffCnt = 0;
30             backoff = false;
31             while (msgIterator.hasNext()) {
32                 int recvColor = Integer.parseInt(msgIterator.next()
33                     .toString());
34                 if (recvColor != myColor) {
35                     usedColors.add(recvColor);
36                 }
37                 // activate backoff
38                 else {
39                     backoff = true;
40                     backOffCnt = (rand.nextInt(lastBO));
41                 }
42             }
43         } else if (backOffCnt > 0) {
44             // decrement backoff
45             backOffCnt--;
46         }
47         // reset color if conflict
48         if (backoff) {
49             backoff = false;
50             myColor = -1;
51         }
52         // process if no backoff
53         if (backOffCnt == 0) {
```

```
54        if (myColor == -1) {
55            for (int i = 0; i < lastColor; i++) {
56                if (!usedColors.contains(i)) {
57                    myColor = i;
58                    super.sendMessageToAllEdges(vertex, new
                         Text(myColor + ""));
59                    break;
60                }
61            }
62        } else {
63            IamDone = true;
64            vertex.setValue(new Text(myColor + ""));
65            vertex.voteToHalt();
66        }
67    }
68  }
69 }
```

## 5.3  Performance Optimizations

In this section, we discuss optimizing Giraph's performance by modifying its runtime parameters. Note that the optimizations in this section depend mainly on the cluster setup, underlying hardware, Giraph's input and algorithm. It may require a custom 3rd party tool or manual tuning through multiple executions to find the optimal configuration values for your Giraph setup.

### 5.3.1  Multithreading

Users can maximize Giraph's performance by increasing the number of threads for each worker. However, there is no golden rule on how many threads are enough to show a positive gain in the performance of Giraph. To avoid overloading Giraph workers, it is a good practice to use the default Giraph setting for multithreading. Expert users, however, can use the following configuration to tune Giraph's performance on Giraph workers with multicore CPUs. The simplest multithreading is to enable parallel computation for Giraph's supersteps. Users can control the number of threads for vertex computation through property giraph.numComputeThreads.

If the input is a huge graph, users can increase Giraph's reading efficiency by increasing the number of threads per worker for reading the input splits. Users can control the number of threads per worker through property giraph.numInput Threads before starting Giraph's job. Users can also use property giraph.use InputSplitLocality to reduce the network overhead by forcing Giraph workers to prioritize reading local input splits before the remote ones, which is enabled by Giraph's default configuration. However, it increases the load on Giraph's ZooKeeper which can significantly affect the performance of reading the input if using a large number of input splits. As a result, it is advised to disable giraph.use

`InputSplitLocality` when using multithreaded input or if the input splits are relatively large.

Similar to input, the output of Giraph can also be multithreaded through property `giraph.numOutputThreads`. Moreover, a single `vertexOutputFormat` can also be shared among parallel `vertexWriters` to parallelize the process of writing the output. This feature is enabled through property `giraph.vertex OutputFormatThreadSafe`, where programmers must ensure that their `giraph.vertexOutputFormatClass` is thread safe to ensure the correctness of the output.

### 5.3.2 Message Exchange Tuning

Giraph supports three different message encoding techniques (`BYTEARRAY_ PER_PARTITION`, `EXTRACT_BYTEARRAY_PER_PARTITION` and `POINTER_ LIST_PER_VERTEX`) that are used to optimize message exchange between workers when there are no message combiners. If a programmer cannot implement message combiners, he should pick the technique that suits best the behavior of his graph algorithm. `BYTEARRAY_PER_PARTITION`, which is the default in Giraph, uses a bit-array to store the messages for each target partition consecutively in memory. This encoding technique uses a 3rd party primitive maps data structure to reduce the total number of used objects. `EXTRACT_BYTEARRAY_PER_PARTITION` is similar to the previous encoding, but it encodes one message for multiple vertices in the same destination partition. This encoding is useful for graph algorithms that broadcast identical messages to a large number of vertices. The third encoding technique is `POINTER_LIST_PER_VERTEX`, where Giraph stores the messages in a byte-array data structure and creates pointers to them for the destination vertices. This encoding storage has a higher overhead in storage but allows an easier access to messages in the byte-array data structure. The message encoding technique is controlled through property `giraph.messageEncodeAndStoreType`.

Giraph does not natively support an enormous number of messages per vertex, i.e., more than 4 Billion messages. Power law graphs usually have such criteria, where a small number of vertices may have edges to lots of other vertices. To remove the space limitation of message encoding classes, users have to assign a `true` value for property `giraph.useBigDataIOForMessages`. When enabling this feature, a vertex may receive messages up to the limitation of the worker's heap space.

### 5.3.3 Controlling the OutEdges Class

In Giraph, four different data structures can be used to store the out-degree edges: ArrayListEdges, ByteArrayEdges, HashMapEdges, HashMultimapEdges. We show in Table 5.1 a comparison between the different data storage classes. To support very large graphs, Giraph uses `ByteArrayEdges` as its default out-degree edges storage. However, users are encouraged to change the out-degree edges storage depend-

**Table 5.1**  A comparison between different out-degree edges storages

| Property | Storage class | | | |
|---|---|---|---|---|
| | ArrayList Edges | ByteArray Edges | HashMap Edges | HashMultimap Edges |
| Memory consumption | Moderate | Low | High | High |
| Edge access | Sequential | Sequential | Random | Random |
| Mutation complexity | Moderate | High | Low | Low |
| Duplicate edges? | Yes | Yes | No | Yes |

ing on their application requirements. For example, if the graph application strictly forbids duplicate edges, programmers are recommended to use `HashMapEdges`. On the other hand, users can use `ArrayListEdges` to balance between storage space and the performance of edge mutations. Users can also implement their own out-degree edges storage by extending `ConfigurableOutEdges` and implementing `MutableOutEdges`. Note that property `giraph.inputOutEdges Class` is used to define out-degree edges storage class when reading from an edge input format, while `giraph.outEdgesClass` is used to generally define Giraph's out-degree edges storage class for all vertices.

### 5.3.4  Out-of-Core Processing

The original design of Giraph is to process big graphs in distributed memory to maximize the performance gain of graph algorithms. By default, Giraph does not handle memory shortage where it relies on the operating system to provide more memory through OS disk swapping. Therefore, Giraph may fail or become too slow if the cluster's memory cannot accommodate the input graph or the messages generated by the user's algorithm. Instead of randomly swapping memory to disk, programmers can utilize Giraph's explicit swapping (out-of-core feature) to efficiently spill least used graph partitions or cache the algorithm's messages temporarily into the disk. To enable out-of-core graph, programmers should assign the value 'true' to property `giraph.useOutOfCoreGraph`, where the maximum number of partitions stored in the memory of each worker is controlled by property `giraph.maxPartitionsInMemory`. Similar to out-of-core graph, programmers enable out-of-core messages by enabling property `giraph.useOutOfCoreMessages` and tune the maximum number of messages stored into disk through property `giraph.maxMessagesInMemory`. The default value for maximum in-memory partitions and in-memory messages in Giraph are 10 partitions and 1 million messages, respectively, where programmers should modify these values based on the memory requirements for each partition and the available memory per worker. Moreover, users have to make sure to direct Giraph to partition the input graph into large number of partitions to ensure that property `giraph.useOutOfCoreGraph` works correctly. For example, if your cluster size is 10 workers and property `giraph.maxPartitionsInMemory` is set to 5, you need to make sure that Giraph initially partitions the input graph more than 50 partitions to enable out-of-core graph processing.

Giraph uses local disks to spill graph partitions and graph messages by using file system(s) assigned by properties `giraph.partitionsDirectory` and `giraph.messagesDirectory`. Programmers can also optimize the buffers used to cache out-of-core messages by modifying property `giraph.messages BufferSize`. Since consistently writing graph partitions into disk is an expensive process, Giraph can be directed to only write the partition's vertices, for static graphs only, in every iteration by assigning the value 'true' to property `giraph.isStaticGraph`. More details can be found in Giraph's website http:// giraph.apache.org/ooc.html. We show in Listing 5.10, an example of using an out-of-core graph and out-of-core messages with 15 partitions and 2 million messages in memory. We also show how to specify two filesystem paths for property `giraph.messagesDirectory`.

Listing 5.10: An example of Giraph's configuration in "giraph-site.xml" to enable out-of-core processing for both partitions and messages

```
1  <configuration>
2  <property>
3    <name>giraph.useOutOfCoreGraph</name>
4    <value>true</value>
5  </property>
6  <property>
7    <name>giraph.maxPartitionsInMemory</name>
8    <value>15</value>
9  </property>
10 <property>
11   <name>giraph.useOutOfCoreMessages</name>
12   <value>true</value>
13 </property>
14 <property>
15   <name>giraph.maxMessagesInMemory</name>
16   <value>2000000</value>
17 </property>
18 <property>
19   <name>giraph.messagesDirectory</name>
20   <value>/mnt/disk2/giraphSpill,
21         /mnt/disk3/giraphSpill</value>
22 </property>
23 </configuration>
```

## 5.4  Giraph Custom Partitioner

While reading the graph, Giraph partitions the graph and distributes its subgraphs to workers on the fly. Giraph uses a deterministic algorithm to bound each vertex to a specific partition while assigning each partition to a particular worker. Giraph provides three partitioning techniques: `HashPartitionerFactory`, `HashRangePartitionerFactory`, `LongMappingStorePartitioner Factory`, `SimpleIntRangePartitionerFactory` and `SimpleLong RangePartitionerFactory`. The default partitioning method is `Hash PartitionerFactory`, where it uses the vertex ID hash to define its partition. This partitioning is the fastest and simplest but does not necessarily guarantee distribution balance between different partitions. `HashRangePartitionerFactory`, on the other hand, provide distribution balance between workers where it uses range partitioning on the hash values of the vertices to distribute them. The other three partitioning techniques (`LongMappingStorePartitionerFactory`, `SimpleIntRangePartitionerFactory` and `SimpleLongRange PartitionerFactory`) are subclasses of the abstract class `Simple Partition`, where they are all based on range partitioning on the actual vertex ID instead of its hash value. Note that all partitioning techniques are based on mathematical calculation to statically assign the location of the vertices and partitions in Giraph. To change the partition class, users can modify property `giraph.graphPartitionerFactoryClass` with either of the above partitioning factories or by using a custom partitioner.

Users can also customize the storage class for each partition, which is controlled by property `giraph.partitionClass`. By default, Giraph uses `Simple Partition` class to store vertices using a map-based data structure. This class allows parallel access to vertices but has higher memory overhead. On the other hand, users may use `ByteArrayPartition` to reduce the partition storage overhead on the cost of parallelism, where only a single thread at a time is allowed to query the vertex information.

## 5.5  Advanced Giraph I/O

Giraph requires one input format class and one output format class to successfully read graphs into memory and write graphs after completing the computation. Through `GiraphConfiguration` class, users can use `setEdgeInputFormatClass` and `setEdgeOutputFormatClass` to assign classes for reading and writing the graphs (Sects. 3.5 and 3.6). Users can either use Giraph's default formats or customize their own to satisfy their application's requirements. To write your personalized format, you have to first identify whether you want to use the adjacency list format or edge list format. Classes that extend `VertexInputFormat`/`VertexOutput Format` are designed to work on inputs/outputs in the format of adjacency lists while extending classes `EdgeInputFormat`/`EdgeOutputFormat` means that your inputs/outputs are in the format of edges lists. Reading and writing in Giraph are independent; you can use any format for Giraph's input and output.

### 5.5.1 Writing a Custom Input Format

To write a custom input format, programmers are required to create a class that implements VertexInputFormat/EdgeInputFormat. However, implementing those classes requires advanced knowledge in Hadoop's I/O. Therefore, we advise users who would like to write their custom input format to extend direct subclasses of VertexInputFormat/EdgeInputFormat, such as TextVertexReader and TextEdgeInputFormat, because they already deal with Hadoop's I/O and only require to know how to translate the input into graph vertices and edges. In Listing 5.11 you can find an example of a simple custom edge reader that extends TextEdgeInputFormat, where our custom class mTextEdgeReaderFrom EachLine in line 6 defines how Giraph should extract the source vertex ID (line 8), destination vertex ID (line 13), and edge value (line 18) from each line from the input file. Note that method createEdgeReader in line 24 is required to allow Giraph to get an instance of our custom class mTextEdgeReaderFromEachLine. Users can also start from a copy from Giraph's implemented input format, such as AdjacencyListTextVertexInputFormat and IntNullTextEdge InputFormat, and modify them as needed based on the user's requirements.

Listing 5.11: Custom edge input format

```
1  public class EdgeInputFormat extends
2  org.apache.giraph.io.formats.TextEdgeInputFormat<LongWritable,
       LongWritable> {
3      public EdgeInputFormat() {
4          super();
5      }
6      protected class mTextEdgeReaderFromEachLine extends
7              TextEdgeReaderFromEachLine {
8          protected LongWritable
               getSourceVertexId(org.apache.hadoop.io.Text line) {
9              return new LongWritable(
10                 Long.parseLong(line.toString().split(":")[0]));
11         }
12
13         protected LongWritable
               getTargetVertexId(org.apache.hadoop.io.Text line) {
14             return new LongWritable(
15                 Long.parseLong(line.toString().split(":")[1]));
16         }
17
18         protected LongWritable getValue(org.apache.hadoop.io.Text
               line) {
19             return new LongWritable(
20                 Long.parseLong(line.toString().split(":")[2]));
21         }
22     }
```

```
23
24    public EdgeReader<LongWritable, LongWritable> createEdgeReader(
25          InputSplit split, TaskAttemptContext context) throws
            IOException {
26       return new mTextEdgeReaderFromEachLine();
27    }
28 }
```

## 5.5.2 Writing a Custom Output Format

Similar to Sect. 5.5.1, users should extend subclasses of `VertexOutputFormat/`
`EdgeOutputFormat`, such as `TextVertexOutputFormat` and `TextEdge`
`OutputFormat`, to implement a custom output format. We show in Listing 5.12 an
example of a custom output format. As previously explained, `TextEdgeWriter`
method in line 17 allows Giraph to get an instance of the user's custom output format class (`mTextEdgeWriterToEachLine`). Note that in this example,
our custom output format class (lines 4–15) only allows edges with positive edge
values to be passed to the output (lines 10–14), where negative edge values are
passed as empty strings. Also note that we are extending a modified version of
`TextEdgeOutputFormat`, which is called `myTextEdgeOutputFormat`. In
this modified class, we modify method `TextEdgeWriterToEachLine.write`
`Edge` (as shown in Listing 13) such that it ignores writing empty strings (lines 5–8)
generated by our `EdgeOutputFormat`.

Listing 5.12: Custom edge output format

```
1  public class EdgeOutputFormat extends
2        myTextEdgeOutputFormat<LongWritable, LongWritable,
          LongWritable> {
3
4     protected class mTextEdgeWriterToEachLine extends
5           TextEdgeWriterToEachLine<LongWritable,
            LongWritable, LongWritable> {
6
7        protected Text convertEdgeToLine(LongWritable sourceId,
8              LongWritable sourceValue, Edge<LongWritable,
                LongWritable> edge)
9              throws IOException {
10          if (edge.getValue().get() < 0) {
11             return new Text("");
12          }
13          return new Text(edge.getValue().get() + ":" +
               sourceValue.get());
14       }
15    }
16
17    public TextEdgeWriter<LongWritable, LongWritable,
          LongWritable> createEdgeWriter(
```

```
18           TaskAttemptContext context) throws IOException,
19           InterruptedException {
20        return new mTextEdgeWriterToEachLine();
21     }
22 }
```

**Listing 5.13: modified `TextEdgeOutputFormat`**

```
1 protected abstract class TextEdgeWriterToEachLine<I extends
      WritableComparable,
2 V extends Writable, E extends Writable> extends
      TextEdgeWriter<I, V, E> { @Override
3    public final void writeEdge(I sourceId, V sourceValue,
        Edge<I, E> edge) throws
4   IOException, InterruptedException {
5       Text tOut = convertEdgeToLine(sourceId, sourceValue,
          edge);
6       if (!tOut.toString().isEmpty()) {
7           getRecordWriter().write(tOut, null);
8       }
9   }
```

## 5.6 Analyzing Giraph Errors

Like any other system, Giraph is prone to runtime errors. Most of the runtime errors are caused by a bug in the user's code, such as `NullPointerException` errors. Other runtime errors may be due to Giraph (or Hadoop) misconfigurations or system bugs. To solve a runtime error, users have to go through the Giraph logs to understand the source of the problem before attempting to solve it. Since Giraph is running on top of Hadoop, Giraph's job logs can be explored in Hadoop's jobtracker web interface (by default found in http://www.hadoopmaster:50030/jobtracker.jsp) and also found in Hadoop's log folder (by default in `HadoopHome/logs/userlogs`). User's code bugs are easy to detect and solve. Configuration errors, on the other hand, are easy to solve, but they require some time to analyze the type or source of the configuration error. System bugs, however, are unavoidable and need community patches to overcome them. You can find the most recent list of Giraph bugs in: https://issues.apache.org/jira/browse/GIRAPH/. Moreover, users can go through Giraph's mailing list http://mail-archives.apache.org/mod_mbox/giraph-user/ to interact with Giraph's community and read about how other users dealt with related Giraph errors. Users are also encouraged to use search engines find possible solutions for the runtime errors. We discuss below the most common configuration errors of Giraph and how to solve them.

### 5.6.1  CountersExceededException

This is one of the most common configuration errors of Giraph, which can occur even when running Giraph's examples. Counters are a global synchronized aggregators in Hadoop. They are originally used by Hadoop to track the MapReduce job statistics, such as the number of Map and Reduce tasks.

```
org.apache.hadoop.mapreduce.counters.LimitExceededException:
    Too many
counters: 121 max=120
```

Giraph uses counters to maintain statistics on each superstep of the user's graph job. For optimization purposes, Hadoop limits the maximum number of counters for each job to 120 counters. Due to this limit, Giraph may fail after a random number of supersteps with showing a `CountersExceededException` in Giraph logs. To avoid this error, users can either increase the value of `mapreduce.job.counters.limit` in Hadoop's configuration file (`mapred-site.xml`) to a higher value or disable superstep counters by assigning a `false` value to `giraph.useSuperstepCounters` in Giraph's configuration.

### 5.6.2  ClassNotFoundException

When building your graph application as a standalone project, you can easily face this error when running Giraph. This error means that Hadoop cannot reach your application class files and, as a result, Giraph cannot start the computation.

```
Exception in thread "main" java.lang.ClassNotFoundException:
    org.apache.giraph.examples.SimpleShortestPathsComputation
```

The recommended way to solve this is to add the path to your project's jar file to the environment variable `CLASSPATH` as follows (in Linux's terminal or in ~/.bashrc):

```
export CLASSPATH=/path/to/jar/file/myCode.jar:$CLASSPATH
```

The other approach to solve the `ClassNotFoundException` is to directly add your project's jar file into Hadoop's `lib` folder. Note that you need to apply the same configuration for all Hadoop servers for either of the above approaches.

### 5.6.3  FileAlreadyExistsException

This error is a common HDFS error, which means that Hadoop's job cannot start because the output path for the job already exists in HDFS.

```
Exception in thread "main"
    org.apache.hadoop.mapred.FileAlreadyExistsException:
Output directory /user/hadoop/output/shortestpaths already
    exists
```

You may face this error if you are writing the graph back into HDFS after the computation of your Giraph's job is finished. To avoid this error, you either delete the output HDFS directory, through command (`hadooop dfs -rmr /user /hadoop/output`), or use another new path as your output path.

### 5.6.4 OutOfMemoryError

This error is common for large graphs when the input graph or the application messages do not fit the allocated memory for any Giraph worker. For servers with big memory, users can increase the allocated memory for Giraph workers on top of Hadoop by modifying property `mapred.child.java.opts` in `mapred-site.xml`.

```
Caused by: java.util.concurrent.ExecutionException:
    java.lang.OutOfMemoryError: Java heap space
  at java.util.concurrent.FutureTask.report(FutureTask.java:122)
  at java.util.concurrent.FutureTask.get(FutureTask.java:202)
  ... 19 more
Caused by: java.lang.OutOfMemoryError: Java heap space
```

In the case that the servers do not have enough memory to accommodate the user's input graph, users must resolve to Giraph's out-of-core processing (Sect. 5.3.4) to avoid `OutOfMemoryError` error.

## 5.7 Failure Recovery

### 5.7.1 Checkpointing

Giraph uses frequent checkpointing to restore the graph computation in the case of any failure. In each checkpoint, the state of the graph is stored in HDFS where the frequency of checkpointing is controlled through property `giraph.checkpoint Frequency`. The default value for this property is 0 which disables Giraph's checkpointing; users can enable checkpointing by assigning an integer value larger than 0 to property `giraph.checkpointFrequency` in Giraph's configuration file 'conf/giraph-site.xml'. This integer value represents the frequency of checkpointing. For example, a value of 7 for property `giraph.checkpointFrequency` in Listing 5.14 means that the checkpointing will occur every 7 supersteps. The best value assignment for the checkpointing frequency depends on the MTTF [51] (Mean Time To Failures) of Hadoop's cluster and Giraph's runtime. Too little checkpoints may lead to expensive failure recovery while too often checkpointing leads to a slower runtime. To increase the efficiency of the checkpointing, programmers can use property `giraph.checkpoint.io.threads` to increase the number of CPU threads for the process of failure recovery. The default number of threads is 8 threads, where programmers should also be aware of the cluster specification to avoid overloading Giraph's workers. An example of how to use property `giraph.checkpoint.io.threads` is shown in Listing 5.14, where we assign 10 threads for the checkpointing process. Checkpointing data can be kept in HDFS after Giraph finishes execution by assigning *false* value to property `giraph.cleanupCheckpointsAfterSuccess`. Moreover, programmers can also control the compression mechanism of checkpointing and change the default HDFS path of checkpointing by modifying properties `giraph.checkpoint.compression.codec` and `giraph.checkpoint Directory`, respectively. More details can be found in Giraph's website: http://giraph.apache.org/options.html. Since checkpointing data is stored in HDFS, we suggest configuring HDFS to have at least 2 dfs replicas to enable HDFS failure recovery. The number of dfs replicas can be set through modifying property `dfs.replication` in Hadoop's configuration file 'hdfs-site.xml'.

Listing 5.14: An example of Giraph's configuration in "giraph-site.xml" to enable checkpointing every 7 supersteps using 10 threads for each worker

```
1  <configuration>
2  <property>
3    <name>giraph.checkpointFrequency</name>
4    <value>7</value>
5  </property>
6  <property>
7    <name>giraph.checkpoint.io.threads</name>
8    <value>10</value>
9  </property>
10 <property>
11   <name>giraph.cleanupCheckpointsAfterSuccess</name>
12   <value>false</value>
13 </property>
14 </configuration>
```

## 5.7.2  Retrying and Recovering Failed Jobs

There are two types of failures: task or worker failures and job failures. If a task or a worker failed, Hadoop would automatically try to recover the work done by that worker if checkpointing was enabled. If checkpointing was disabled, Hadoop would not recover the failed task (Hadoop Mapper) and Giraph will consider that the whole job has failed. If a job fails, by default Giraph does not try to restart or recover it. This behavior is controlled through property `giraph.jobRetryCheckerClass`, where its default class is `DefaultGiraphJobRetryChecker`. Users can implement interface `GiraphJobRetryChecker` to change the behavior Giraph's job retry and restart policies. Interface `GiraphJobRetryChecker` has two functions: `shouldRetry()` and `shouldRestartCheckpoint()`. If a job failed, Giraph first invokes function `shouldRestartCheckpoint()` to ask the user if this job should restart from the latest checkpoint. If there is not a suitable checkpoint or the user decided not to use the checkpoint, Giraph then invokes function `shouldRetry()`. This function asks the user, given the current number of retries, if he would like to retry the whole job from the beginning or not.

# Related Large-Scale Graph Processing Systems

**6**

In practice, the introduction of Google's Pregel system followed by Apache Giraph, as its open source realization, has inspired the development of other various large-scale graph processing systems. In this chapter, we highlight and provide an overview of two other popular systems in this domain, namely, **GraphX** and **GraphLab**.

## 6.1 GraphX

### 6.1.1 Spark and RDD

Apache Spark is a cluster computing platform designed to be fast and general purpose [52]. Spark, written in Scala, was originally developed in the AMPLab at UC Berkeley and open-sourced in 2010 as one of the new generation data flow engines that follows in the line of the MapReduce framework. In particular, while Hadoop introduced a radical new approach which is based on distributing the data when it is stored and running the computation where the data resides, however, one of the main limitation of the Hadoop framework is that it requires that the entire output of each map and reduce task to be materialized into a local file on the Hadoop Distributed File System (HDFS) [53] before it can be consumed by the next stage. This materialization step facilitates the implementation of a simple and elegant check-point/restart fault tolerance mechanism, however, it dramatically harms the system performance. Spark takes the concepts of Hadoop to the next level by loading the data in distributed memory and relying on less expensive shuffles during data processing. Figure 6.1 contrasts the main difference between the Spark framework and the Hadoop framework.

Spark combines an execution engine for distributing programs across clusters of machines with an elegant and rich programming model. In particular, the funda-

© Springer International Publishing AG 2016
S. Sakr et al., *Large-Scale Graph Processing Using Apache Giraph*,
DOI 10.1007/978-3-319-47431-1_6

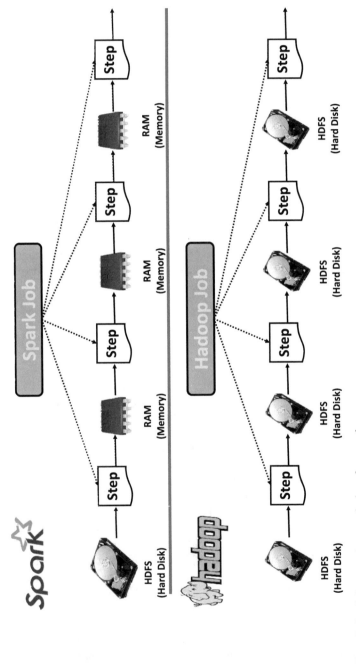

**Fig. 6.1** Spark framework versus hadoop framework

mental programming abstraction of Spark is called Resilient Distributed Datasets (RDD), an in-memory data structure which leverages the power of Spark's functional programming paradigm by allowing user programs to load data into a cluster's memory and query it repeatedly. In principle, the RDD abstraction enables developers to materialize any point in a processing pipeline into memory across the cluster, meaning that future steps that want to deal with the same dataset need not recompute it or reload it from disk. In addition, developers can explicitly cache an RDD in memory across machines and reuse it in multiple parallel operations. Thus, Spark is well suited for highly iterative algorithms that require multiple passes over a dataset, as well as reactive applications that quickly respond to user queries by scanning large in-memory datasets. This has been considered one of the main bottleneck scenarios for the Hadoop framework.

In Spark, RDD is an immutable data structure which means that data in an RDD is never changed, distributed collection of objects. In principle, an RDD is laid out across the cluster of machines as a collection of *partitions*, each including a subset of the data. The partitioning process is done automatically by Spark, however, optionally, developers can control how many partitions should be used for an RDD. Partitions define the unit of parallelism in Spark. During data processing, Spark processes the objects within a partition in sequence, and processes multiple partitions in parallel. RDDs can hold any type of element including primitive types (e.g., integers, characters, boolean, etc.), sequence types (e.g., strings, lists, arrays, tuples, etc.), nested data types, Scala/Java Objects (if serializable) in addition to mixed types. A specific form of RDDs is *Pair RDD* where each element of the RDD must be a key-value pair. In practice, developers can create RDDs in various way. For example, RDD can be created by loading data from a file or a set of files where each line in the file (s) is represented as a separate record in the RDD. The following code snippet shows an example of loading file(s) into RDD using the *SparkContext*:

```
//single file
sc.textFile("mydata.txt")
//comma-separated list of files
sc.textFile("mydatafile1.txt, mydatafile2.txt")
// a wildcare list of file
sc.textFile("mydirectory/*.txt")
```

RDD can be also created by distributing a collection of objects (e.g., a list or set) which are loaded in memory or by applying coarse grained transformations (e.g., map, filter, reduce, join) on existing RDDs. Spark depends heavily on the concepts of functional programming. In Spark, functions represent the fundamental unit of programming where functions can only have input and output but with no state or side effects. In principle, Spark offers two types of operations over RDDs: *transformations* and *actions*. Transformations are used to construct a new RDD from an existing one. For example, one common transformation is filtering data that matches a predicate. On the other hand, Actions is used to compute a result based on an existing RDD, and return the results either to the driver program or save it to an external storage system (e.g., HDFS). Figure 6.2 illustrates the flow of RDD operations in Spark. Table 6.1

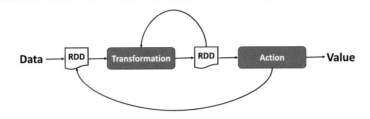

Fig. 6.2: Flow of RDD operations in spark

Table 6.1: Sample spark's transformations

| Transformation | Description |
|---|---|
| map | Apply a transformation function to each element in the input RDD and return a new RDD with the elements of the transformation output as a result |
| filter | Apply a filtration predicate on the elements of an RDD and returns a new RDD with only the elements which satisfy the predicate conditions |
| distinct | Remove the duplicate elements of an RDD |
| union | Return all elements of two RDDs |
| cartesian | Return the cartesian product of the elements of two RDDs |
| intersection | Return the elements which are contained in two RDDs |
| subtract | Return the elements which are not contained in another RDD |

Table 6.2: Sample spark's actions

| Action | Description |
|---|---|
| take | Return number of elements from an RDD |
| takeOrdered | Return number of elements from an RDD based on defined order |
| top | Return the top number of elements from an RDD |
| count | Return the number of elements in an RDD |
| countByValye | Return the number of times each element occurs in an RDD |
| reduce | Combine the elements on an RDD together according to an aggregate function |
| foreach | Apply a function for each element in an RDD |

provides an overview of some Spark's transformation operations while Table 6.2 provides an overview of some Spark's action operations.

The Resilient feature of an RDD means that if data in memory is lost; it can be recreated. In particular, RDDs achieve fault tolerance through a notion of lineage so that a resilient distributed dataset (RDD) can be rebuilt if a partition is lost. In other words, instead of relying on schemes for persisting or check-pointing intermediate results, Spark remembers the sequence of operations which led to a certain data set. In particular, if a partition of an RDD is lost, the RDD has enough information about how it was derived from other RDDs to be able to rebuild just that partition. For cluster management, Spark supports both of Apache Mesos [54] and Hadoop YARN [55].

Spark's core APIs are designed to cover a wide range of workloads that previously required separate distributed systems, including batch applications, iterative algorithms, interactive queries and streaming. The Spark Shell provides an interactive data exploration (REPL) environment. In addition, Spark supports various APIs in Python, Java, Scala and provides rich built-in libraries. In practice, one of the common scenarios is that users need to be able to break down the barriers between data silos such that they are able to design computations or analytics that combine different types of data (e.g., structured, unstructured, stream, graph). In order to achieve this goal, Spark provides various packages with higher level libraries including support for SQL queries (Spark SQL) [56], streaming data (Spark Streaming),[1] machine learning (MLib) [57], statistical programming (Spark R[2]) and graph processing (GraphX) [58]. These libraries increase developer productivity and can be seamlessly combined to create complex workflows. In addition, it reduces the management burden of maintaining separate tools. Figure 6.3 provides an overview of Spark's ecosystem. In the following, we provide an overview of the main Graph interface which is provided by Spark, GraphX.

## 6.1.2   GraphX RGD

GraphX[3] is a distributed graph engine built on top of Spark [58]. GraphX extends Spark's Resilient Distributed Dataset (RDD) abstraction to introduce the Resilient Distributed Graph (RDG), which associates records with vertices and edges in a graph and provides a collection of expressive computational primitives. The GraphX RDG leverages advances in distributed graph representation and exploits the graph structure to minimize communication and storage overhead. It relies on a flexible vertex-cut partitioning to encode graphs as horizontally partitioned collections. In contrast to other graph systems (e.g., Pregel and Giraph), RDG represents graph as records in tabular views of vertices and edges. In particular, as illustrated in Fig. 6.4, in RDG, a graph is represented as directed adjacency structure which is mapped into tabluar records. Similar to RDDs, in RDGs, transformation from one graph to the next creates another RDG with transformed vertices and edges via transformation operators. While the basic GraphX RDG interface naturally expresses graph transformations, filtering operations, and queries, it does not directly provide an API for recursive graph-parallel algorithms. Instead, the GraphX interface is designed to enable the construction of new graph-parallel APIs. In addition, unlike other graph processing systems, the GraphX API enables the composition of graphs

---

[1]http://www.spark.apache.org/streaming/.
[2]https://www.spark.apache.org/docs/1.6.0/sparkr.html.
[3]http://www.spark.apache.org/graphx/.

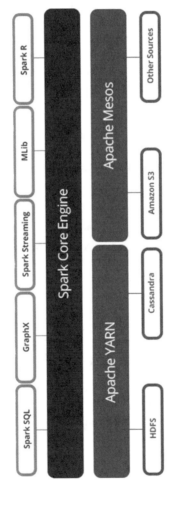

**Fig. 6.3** Spark's ecosystem

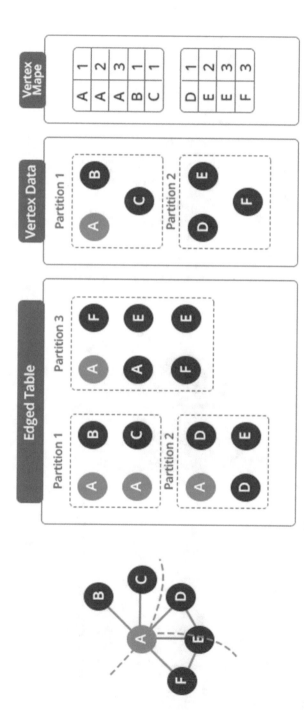

**Fig. 6.4** RDG representation of graphs in graphX

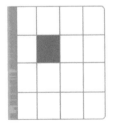

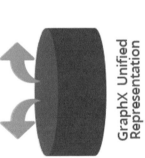

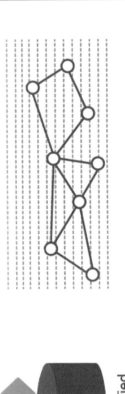

**Fig. 6.5** Unified representation of graphs in graphX

with unstructured and tabular data and allows the same physical data to be viewed both as a graph and as collections without data movement or duplication (Fig. 6.5).

GraphX also provides various operators for manipulating graphs (e.g., subgraph and mapVertices) and a library of common graph algorithms (e.g., PageRank and triangle counting). GraphX reuses indices across graph and collection views and over multiple iterations, thus reducing memory overhead and improving system performance. By leveraging logical partitioning and lineage, GraphX achieves low-cost fault tolerance. In addition, by exploiting immutability, GraphX reuses indices across graph and collection views and over multiple iterations, reducing memory overhead and improving system performance.

## 6.1.3 Examples

Assume we want to compute the connected components in a graph of heterogeneous collection of text sources. The text source can be a social network post, an article in a magazine, a page from a web site, a books or a scientific paper. Each text source contains a set of words, one or more subjects and may reference another text source. Such a problem is very common, and it can be solved in many different ways. Since we are presenting a graph processing system in this book, we will approach this problem as a graph problem.

The first step to solve this problem as a graph problem is to represent different text sources as a directed graph. This graph is going to have three different types of vertices: a word vertex ($W$), a text source vertex ($T$), and a subject vertex ($S$).

Each unique word, in the domain of text sources, is represented by an independent vertex of type $W$. Moreover, each independent text source and its subject(s) are also represented by a $T$ vertex and an S vertex, respectively. An edge is added between $W_i$ and $T_j$ vertices whenever a text source $T_j$ contains the word represented by vertex $W_i$. When a text source contains a word, an edge is added between the $W$ vertex of that word and the $T$ vertex of the text source. Moreover, an edge is also added between two $T$ vertices when one text source cites another. Note that all edges are weighted, where each weight represents the frequency of that edge between different vertices. We show in Fig. 6.6 an example graph of a word-text-subject graph.

**Listing 6.1: Example text snippet**

```
1 A camel is an even-toed ungulate within the genus Camelus,
      bearing
2 distinctive fatty deposits known as humps on its back. The
      two
3 surviving species of camel are the dromedary and the
      bactrian.
4
5 ==Further reading==
6 *cite book|last=Gilchrist|first=W.|year=1851|title=A
      Practical Treatise
```

```
 7│ on the Treatment of the Diseases of the Elephant, Camel &
       Horned Cattle:
 8│ with instructions for improving their efficiency; also, a
       description of
 9│ the medicines used in the treatment of their diseases; and a
       general
10│ outline of their
       anatomy'|location=Calcutta|publisher=Military Orphan
11│ Press
```

The first step in converting the text sources into a graph, we have to change its free text style into formatted text with informative content in each line. In this context, we assume all text sources can be loaded into memory as strings. For example, we convert a snippet of the Wikipedia page on Camel, as shown in Listing 6.1, into a formatted text, as shown in in Listing 6.2 where each line contains the text source type, the related information (such as subject, word, or another text source) and

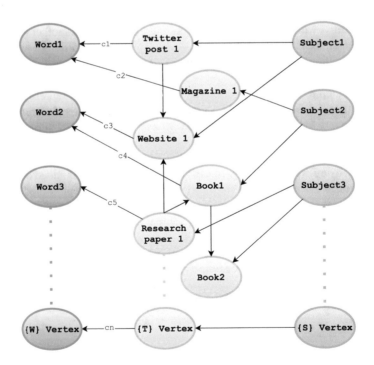

Fig. 6.6: Word-text-subject graph

the weight of the edge. For example, Line 1 in Listing 6.2 has type S that denotes a subject entry, Wikipedia:Camel for the text source, Camel for the subject and 1 for the weight. Moreover, Lines 2–32 contains the words in the Wikipedia text source while Line 33 contains the citation to another text source (identified by Book:B00089UG52). Note that Listing 6.2 is just an example of how to format free text into meaningful readable entries, however, users are free to use whatever method and annotations available to achieve similar results.

**Listing 6.2: Formatted text snippet**

```
 1 S Wikipedia:Camel Camel     1
 2 W Wikipedia:Camel camel     1
 3 W Wikipedia:Camel is        1
 4 W Wikipedia:Camel an        1
 5 W Wikipedia:Camel even-toed  1
 6 W Wikipedia:Camel ungulate  1
 7 W Wikipedia:Camel within    1
 8 W Wikipedia:Camel the       1
 9 W Wikipedia:Camel genus     1
10 W Wikipedia:Camel Camelus   1
11 W Wikipedia:Camel bearing   1
12 W Wikipedia:Camel distinctive  1
13 W Wikipedia:Camel fatty     1
14 W Wikipedia:Camel deposits  1
15 W Wikipedia:Camel known     1
16 W Wikipedia:Camel as        1
17 W Wikipedia:Camel humps     1
18 W Wikipedia:Camel on        1
19 W Wikipedia:Camel its       1
20 W Wikipedia:Camel back      1
21 W Wikipedia:Camel the       1
22 W Wikipedia:Camel two       1
23 W Wikipedia:Camel surviving  1
24 W Wikipedia:Camel species   1
25 W Wikipedia:Camel of        1
26 W Wikipedia:Camel camel     1
27 W Wikipedia:Camel are       1
28 W Wikipedia:Camel the       1
29 W Wikipedia:Camel dromedary  1
30 W Wikipedia:Camel and       1
31 W Wikipedia:Camel the       1
32 W Wikipedia:Camel bactrian  1
33 T Wikipedia:Camel Book:B00089UG52 1
```

After formatting the text source, we use SparkContext.textFile(), Line 4 in Listing 6.5, to load from a single text file into one RDD. If a user has multiple input text files in a single directory, he can use SparkContext.wholeTextFiles() to load multiple text files into a single RDD. After loading the input into an RDD, we convert the input into an array of strings, by using the writable class TextArrayWritable in Listing 6.4, to separate the different entities of the input

String (Lines 7–15 in Listing 6.5). After that, we adjust the weights of edges by combining identical entries and increase their weights. This is done by identifying a unique key for each RDD entry (Lines 18–24 in Listing 6.5), and then using the reduce command (Lines 27–38 in Listing 6.5) to combine the weight of identical entries. We show in Listing 6.3 the result of the reduce operation.

**Listing 6.3: Formatted text snippet while combining identical entries**

```
 1 S Wikipedia:Camel Camel    1
 2 W Wikipedia:Camel camel    2
 3 W Wikipedia:Camel is       1
 4 W Wikipedia:Camel an       1
 5 W Wikipedia:Camel even-toed 1
 6 W Wikipedia:Camel ungulate 1
 7 W Wikipedia:Camel within   1
 8 W Wikipedia:Camel the      4
 9 W Wikipedia:Camel genus    1
10 W Wikipedia:Camel Camelus  1
11 W Wikipedia:Camel bearing  1
12 W Wikipedia:Camel distinctive 1
13 W Wikipedia:Camel fatty    1
14 W Wikipedia:Camel deposits 1
15 W Wikipedia:Camel known    1
16 W Wikipedia:Camel as       1
17 W Wikipedia:Camel humps    1
18 W Wikipedia:Camel on       1
19 W Wikipedia:Camel its      1
20 W Wikipedia:Camel back     1
21 W Wikipedia:Camel two      1
22 W Wikipedia:Camel surviving 1
23 W Wikipedia:Camel species  1
24 W Wikipedia:Camel of       1
25 W Wikipedia:Camel are      1
26 W Wikipedia:Camel dromedary 1
27 W Wikipedia:Camel and      1
28 W Wikipedia:Camel bactrian 1
29 T Wikipedia:Camel Book:B00089UG52 1
```

**Listing 6.4: Custom ArrayWritable class**

```
 1 public static class TextArrayWritable extends ArrayWritable {
 2     public TextArrayWritable() {
 3       super(Text.class);
 4     }
 5
 6     public TextArrayWritable(String[] strings) {
 7       super(Text.class);
 8       Text[] texts = new Text[strings.length];
 9       for (int i = 0; i < strings.length; i++) {
10         texts[i] = new Text(strings[i]);
```

```
11        }
12      set(texts);
13    }
14  }
```

**Listing 6.5: Reading a file from disk and converting it into a list of TextArrayWritable**

```
1  String wikiPage = "/tmp/wikiCamel.txt";
2
3  //Loading into RDD with minimum number of partitions == 1
4  JavaRDD<String> input = sc.textFile(wikiPage,1);
5
6  //converting the input into TextArrayWritable
7  JavaRDD<TextArrayWritable> formattedInput = input
8          .map(input -> {
9            String[] output = new String[4];
10           output[0] = input.split("")[0];
11           output[1] = input.split("")[1];
12           output[2] = input.split("")[2];
13           output[3] = input.split("")[3];
14           return new TextArrayWritable(output);
15         });
16
17  //Adding a key for each value in the RDD
18  JavaPairRDD<LongWritable, TextArrayWritable> edgeKeys =
        formattedInput
19          .keyBy(input -> {
20            Text[] myInput = (Text[]) input.get();
21            return new LongWritable((myInput[0].toString() + "-"
22              + myInput[1].toString() + "-" + myInput[2]
23              .toString()).hashCode());
24          });
25
26  //Combine similar edges
27  JavaRDD<TextArrayWritable> uniqueEdges =
        edgeKeys.reduceByKey(
28          (input1, input2) -> {
29            Text[] myInput = (Text[]) input1.get();
30            String[] output = new String[4];
31            output[0] = myInput[0].toString();
32            output[1] = myInput[1].toString();
33            output[2] = myInput[2].toString();
34            int cnt1 = Integer.parseInt(myInput[3].toString());
35            int cnt2 =
                  Integer.parseInt(input2.get()[3].toString());
36            output[3] = ((cnt1 + cnt2) + "");
37            return new TextArrayWritable(output);
38          }).values();
```

Next, we convert the RDD into a graph format for GraphX. We first generate a list of all unique vertices and then convert the edges into RDDs of `Edge` classes. To create unique vertices, we read each edge and create a list of its source and destination vertices. Each vertex is identified by the hash code of its type and value. After that, we reduce the RDD of vertices based on the vertex IDs to remove duplicate vertices. We show a snippet code of this operation in Listing 6.6.

**Listing 6.6: Extracting unique edges**

```
1  //Creating unique graph vertices
2  JavaPairRDD<LongWritable, Text> vertices = uniqueEdges
3          .flatMapToPair(input -> {
4              Text[] myInput = (Text[]) input.get();
5              ArrayList<Tuple2<LongWritable, Text$\,\gg\,$output =
                  null;
6              String v1 = "T-" + myInput[1].toString();
7              String v2 = myInput[0].toString() + "-"
8                  + myInput[2].toString();
9              output.add(new Tuple2<>(new
                  LongWritable(v1.hashCode()),
10                 new Text(v1)));
11             output.add(new Tuple2<>(new
                  LongWritable(v2.hashCode()),
12                 new Text(v2)));
13             return output;
14         });
15
16 JavaPairRDD<LongWritable, Text> uniqueVertices =
       vertices.reduceByKey((
17         input1, input2) -> input1)
```

Next, we convert the unique edges into an RDD of class `Edge`. This is done by converting each entry in the RDD to an RDD of edges by using a map function on the text source RDD (Lines 2–9 in Listing 6.7). If a user generated multiple edge RDDs, he can join them by using `RDD.union(another RDD)`. At this stage, the graph is fully constructed and ready for graph processing. Assuming that the vertex unique ID is used as a vertex value, the constructed graph should be similar to the example graph in Fig. 6.7. To start the graph processing stage, we pass the RDD of edges to GraphX to apply our connected component. Since there are no Java API for GraphX, we use Scala to pass the RDD of edges to GraphX as shown in Listing 6.8.

**Listing 6.7: Converting into GraphX format and processing the graph**

```
1  //Converting from String RDD into
2  Edge RDD JavaRDD<Edge<LongWritable$\,\gg\,$graphEdges =
       uniqueEdges.map(input -> {
3          Text[] myInput = (Text[]) input.get();
4          return new Edge<LongWritable>(
5              (myInput[0].toString() + "-" +
                  myInput[2].toString())
```

```
6        .hashCode(), ("T-" + myInput[1].toString())
7        .hashCode(), new LongWritable(Long
8        .parseLong(myInput[3].toString()))));
9  });
10
11 //Building the graph in GraphX, and compute our algorithms
12 JavaRDD<EdgeTriplet<Object, LongWritable$\,\gg\,$result = new
       RDDGraphBuilder(
13         edges.rdd()).buildGraphRDD();
```

**Listing 6.8: Passing JavaRDD<Edge> to GraphX**

```
1 class RDDGraphBuilder(var edgeRDD: RDD[Edge[LongWritable]]) {
2   def buildGraphRDD: JavaRDD[EdgeTriplet[VertexId,
      LongWritable]] = {
3     var g: Graph[Integer, LongWritable] =
        Graph.fromEdges(edgeRDD, 0)
4     //Assuming GraphX process the graph as undirected when
        finding connectedComponents
5     new JavaRDD[EdgeTriplet[VertexId,
        LongWritable]](g.connectedComponents().triplets)
6   }
7 }
```

The result is returned in line 12 of Listing 6.7 contains the connected components IDs for the source and destination vertices in each edge of the graph. In this example, we assume that the connected component in GraphX is processed in the input graph as an undirected graph. For example, the graph in Fig. 6.8 represents the result of GraphX connected components on the graph in Fig. 6.7. Note that the actual vertex IDs are simply hashes of the original vertex text value. To recover the original content of the vertices, users can join this output with the unique vertices RDD in Line 16 of Listing 6.6. The connected components is useful to identify independent subgraphs. For example, both vertices Travel and Banana boat in Fig. 6.8 exists in the same subgraph because both vertices have a value 1. Moreover, vertices Result and Book:SeaSports are not in the same subgraph because they have different component IDs.

## 6.2 GraphLab

### 6.2.1 Programming Model

*GraphLab* [59] is an open-source large-scale graph processing project, implemented in C++, which started at CMU and is currently supported by Dato Inc.[4] Unlike

---

[4]https://www.dato.com/.

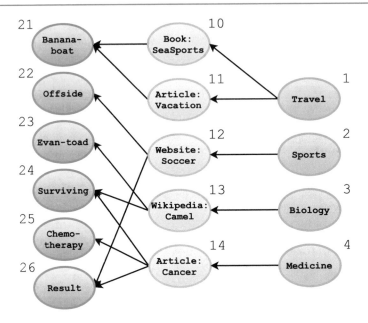

Fig. 6.7: An example graph with vertex IDs used as vertex values

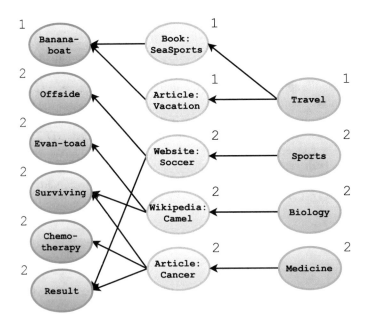

Fig. 6.8: The result of undirected connected components on the example graph in Fig. 6.7 with vertex IDs used as vertex values

Pregel, GraphLab relies on the shared memory abstraction and the GAS (Gather, Apply, Scatter) processing model which is similar to but also fundamentally different from the BSP model that is employed by Pregel. The GraphLab abstraction consists of three main parts: the *data graph*, the *update function*, and the *sync operation*. The data graph represents a user-modifiable program state that both stores the mutable user-defined data, and encodes the sparse computational dependencies. The update function represents the user computation and operates on the data graph by transforming data in small overlapping contexts called *scopes*. In the GAS model, a vertex collects information about its neighborhood in the *Gather* phase, performs the computations in the *Apply* phase, and updates its adjacent vertices and edges in the *Scatter* phase. As a result, in GraphLab, graph vertices can directly *pull* their neighbours' data (via Gather) without the need to explicitly receive messages from those neighbors. In contrast, in the BSP model of Pregel, a vertex can learn its neighbors' values only via the messages that its neighbors *push* to it. GraphLab offers two execution modes: *synchronous* and *asynchronous*. Like BSP, the synchronous mode uses the notion of communication barriers while the asynchronous mode does not support the notion of communication barriers or supersteps. It uses distributed locking to avoid conflicts and to maintain serializability. In particular, GraphLab automatically enforces serializability by preventing adjacent vertex programs from running concurrently by using a fine-grained locking protocol that requires sequentially grabbing locks on all neighboring vertices. For instance, it uses an *Edge Consistency Model* that allows two vertices to be simultaneously updated if they do not share an edge. In addition, it applies a *Graph Coloring* mechanism where two vertices can be assigned the same color if they do not share an edge. It is the jobs of the system the scheduler to determine the order that vertices can be updated.

In practice, high-degree vertices in power-law graphs causes the workload to be imbalanced during the execution of graph computations. Therefore, another member of the GraphLab family of systems, *PowerGraph* [60], has been introduced to tackle this challenge. In particular, PowerGraph relies on a vertex-cut partitioning scheme (Fig. 6.9) that cuts the vertex set in such a way that the edges of a high-degree vertex are handled by multiple workers. Therefore, as a tradeoff, vertices are replicated across workers, and communication among workers are required to guarantee that the vertex value on each replica remains consistent. PowerGraph eliminates the degree dependence of the vertex program by directly exploiting the GAS decomposition to factor vertex programs over edges. Therefore, it is able to retain the "think-like-a-vertex" programming style while distributing the computation of a single vertex program over the entire cluster. In principle, PowerGraph attempts to merge the best features from both Pregel and GraphLab. From GraphLab, PowerGraph inherits the data graph and shared-memory view of computation thus eliminating the need for users to specify the communication of information. From Pregel, PowerGraph borrows the commutative, associative gather concept. PowerGraph supports both the highly parallel bulk-synchronous Pregel model of computation as well as the computationally efficient asynchronous GraphLab model of computation.

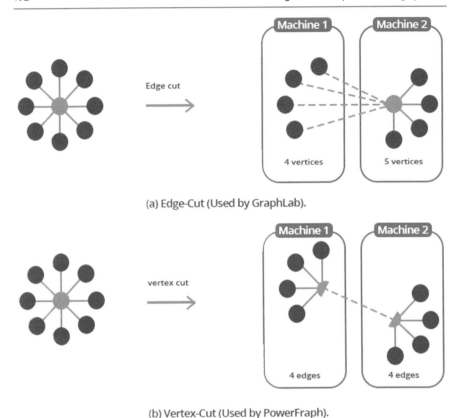

(a) Edge-Cut (Used by GraphLab).

(b) Vertex-Cut (Used by PowerFraph).

Fig. 6.9: Edge-Cut versus vertex-cut partitioning schemes

Another member of the GraphLab family of systems is *GraphChi* [61]. Unlike
the other *distributed* members of the family, GraphChi,[5] implemented in C++, is
a *centralized* system that can process massive graphs from secondary storage in a
single machine. In particular, GraphChi relies on a *Parallel Sliding Windows* (PSW)
mechanism for processing very large graphs from disk. PSW is designed to require
only a very small number of nonsequential accesses to the disk, and thus it can
perform well on both SSDs and traditional hard drives. PSW partitions the input
graph into subgraphs, called shards. In each shard, edges are sorted by the source
IDs and loaded into memory sequentially. In addition, GraphChi supports a selective
scheduling mechanism that attempts to converge faster on some parts of the graph
especially on those where the change on values is significant. The main advantage
of systems like GraphChi is that it avoids the challenge of finding efficient graph
cuts that are balanced and can minimize the communication between the workers,

---

[5]http://www.graphlab.org/projects/graphchi.html.

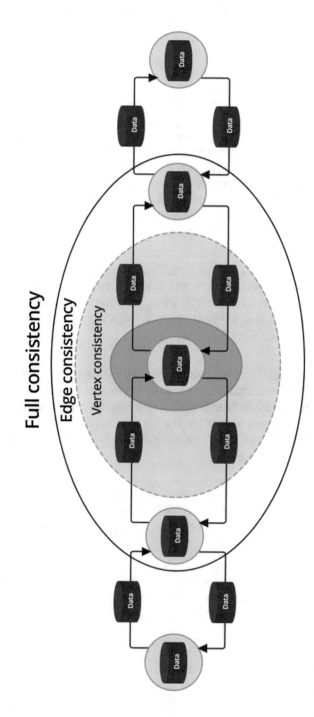

**Fig. 6.10** GraphLab consistency models

which is a hard challenge. It also avoids other challenges of distributed systems such as cluster management and fault tolerance.

## 6.2.2 Consistency Model

To ensure race-free computation in asynchronous algorithms, updates should be consistent. It is a generally thought that forgoing consistency in statistical algorithms does not affect much but many algorithms require strict consistency or perform significantly better under strict consistency. Figure 6.10 shows the number of iterations the Alternating Least Squares algorithm takes to converge using inconsistent and consistent updates. This shows that if consistency is ensured, some algorithms converge very fast. GraphLab ensures sequential consistency which means that for each parallel execution, there exists a sequential execution of update functions which produces the same result. GraphLab supports following consistency models.

**Full Consistency**
In this model, consistency is ensured in the complete scope of a vertex. This means that when the update function of a vertex is being executed, only this vertex will have a read–write access to the data of its neighboring vertices and edges. When this vertex is being executed, the vertices whose scope overlaps its scope will not be executed until this vertex does not finish.

**Edge Consistency**
Full consistency model makes the algorithm run slow because fewer vertices can run simultaneously. For those algorithms which do not require full consistency, another model is supported which is called Edge Consistency. In this model, consistency is ensured for a vertex's own data as well as the data of its neighboring edges. This relaxes the consistency and allows more vertices to run their update functions in parallel.

**Vertex Consistency**
GraphLab also supports a more relaxed consistency model in which consistency is ensured at vertex level. When a vertex is being executed, no other vertex can have access to its data. This greatly enhances the speed of algorithms which require consistency only on the vertices.

**No Consistency**
The consistency can also be switched off. This means that the scheduler is free to schedule the execution of any vertex that is available. This offers more parallelism and is faster than the three consistency models mentioned before.

# References

1. Large synoptic survey. http://www.lsst.org/
2. Hey, T., Tansley, S., Tolle, K. (eds.): The Fourth Paradigm: Data-Intensive Scientific Discovery. Microsoft Research (2009)
3. Bell, G., Gray, J., Szalay, A.S.: Petascale computational systems. IEEE Comput. **39**(1), 110–112 (2006)
4. Manyika, J., Chui, M., Brown, B., Bughin, J., Dobbs, R., Roxburgh, C., Byers, A.H.: Big data: the next frontier for innovation, competition, and productivity. Technical report 1999–66 (2011)
5. Dean, J., Ghemawa, S.: MapReduce: simplified data processing on large clusters. In: OSDI (2004)
6. Yang, H., Dasdan, A., Hsiao, R., Parker, D.: Map-reduce-merge: simplified relational data processing on large clusters. In *SIGMOD* (2007)
7. Stonebraker, M.: The case for shared nothing. IEEE Database Eng. Bull. **9**(1), 4–9 (1986)
8. White, T.: Hadoop: The Definitive Guide. O'Reilly Media (2012)
9. Sakr, S., Liu, A., Fayoumi, A.G.: The family of mapreduce and large-scale data processing systems. ACM Comput. Surv. **46**(1) (2013)
10. Sakr, S., Pardede, E. (eds.): Graph Data Management: Techniques and Applications. IGI Global (2011)
11. Khan, A., Elnikety, S.: Systems for big-graphs. PVLDB **7**(13), 1709–1710 (2014)
12. O'reilly, T.: What is web 2.0: design patterns and business models for the next generation of software. Commun. Strateg. (1):17 (2007)
13. Scott, J.: Social Network Analysis. Sage (2012)
14. Linden, G., Smith, B., York, J.: Amazon. com recommendations: item-to-item collaborative filtering. IEEE Internet Comput. **7**(1), 76–80 (2003)
15. Broder, A., Kumar, R., Maghoul, F., Raghavan, P., Rajagopalan, S., Stata, R., Tomkins, A., Wiener, J.: Graph structure in the web. Comput. Netw. **33**(1), 309–320 (2000)
16. Page, L., Brin, S., Motwani, R., Winograd, T.: The pagerank citation ranking: bringing order to the web (1999)
17. Trochim, W.M.K., Donnelly, J.P.: Research methods knowledge base (2001)
18. Bizer, C., Heath, T., Berners-Lee, T.: Linked data-the story so far. In: Semantic Services, Interoperability and Web Applications: Emerging Concepts, pp. 205–227 (2009)

© Springer International Publishing AG 2016

S. Sakr et al., *Large-Scale Graph Processing Using Apache Giraph*,
DOI 10.1007/978-3-319-47431-1

19. Auer, S., Bizer, C., Kobilarov, G., Lehmann, J., Cyganiak, R., Ives, Z.: Dbpedia: A Nucleus for a Web of Open Data. Springer, Berlin (2007)
20. Urbani, J., Kotoulas, S., Oren, E., Van Harmelen, F.: Scalable Distributed Reasoning Using Mapreduce. Springer (2009)
21. Steenstra, J., Gantman, A., Taylor, K., Chen, L.: Location based service (lBS) system and method for targeted advertising. US Patent App. 10/931, 309 (2004)
22. Caragliu, A., Del Bo, C., Nijkamp, P.: Smart cities in Europe. J. Urban Technol. **18**(2), 65–82 (2011)
23. Zheng, Y., Chen, Y., Xie, X., Ma, W.-Y.: Geolife2.0: a location-based social networking service. In: Tenth International Conference on Mobile Data Management: Systems, Services and Middleware, 2009. MDM'09, pp. 357–358. IEEE (2009)
24. Trinajstic, N., et al.: Chemical Graph Theory. CRC Press (1992)
25. Jeong, H., Mason, S.P., Barabási, A.-L., Oltvai, Z.N.: Lethality and centrality in protein networks. Nature **411**(6833), 41–42 (2001)
26. Robinson, I., Webber, J., Eifrem, E.: Graph Databases: New Opportunities for Connected Data. O'Reilly Media (2015)
27. Partner, J., Vukotic, A., Watt, N., Abedrabbo, T., Fox, D.: Neo4j in Action. Manning Publications Company (2014)
28. Martinez-Bazan, N., Gomez-Villamor, S., Escale-Claveras, F.: Dex: A high-performance graph database management system. In: 2011 IEEE 27th International Conference on Data Engineering Workshops (ICDEW), pp. 124–127. IEEE (2011)
29. Iordanov, B.: Hypergraphdb: a generalized graph database. In: Web-Age Information Management, pp. 25–36. Springer (2010)
30. Bu, Y., Howe, B., Balazinska, M., Ernst, M.D.: The HaLoop approach to large-scale iterative data analysis. VLDB J. **21**(2) (2012)
31. Ekanayake, J., Li, H., Zhang, B., Gunarathne, T., Bae, S.-H., Qiu, J., Fox, G.: Twister: a runtime for iterative MapReduce. In: HPDC (2010)
32. Zhang, Y., Gao, Q., Gao, L., Wang, C.: iMapReduce: a distributed computing framework for iterative computation. J. Grid Comput. **10**(1) (2012)
33. Chen, R., Weng, X., He, B., Yang, M.: Large graph processing in the cloud. In: SIGMOD (2010)
34. Kang, U., Tsourakakis, C.E., Faloutsos, C.: PEGASUS: a peta-scale graph mining system. In: ICDM (2009)
35. Kang, U., Tong, H., Sun, J., Lin, C.-Y., Faloutsos, C.: GBASE: a scalable and general graph management system. In: KDD (2011)
36. Kang, U., Tsourakakis, C.E., Faloutsos, C.: PEGASUS: mining peta-scale graphs. Knowl. Inf. Syst. **27**(2) (2011)
37. Kang, U., Meeder, B., Faloutsos, C.: Spectral analysis for billion-scale graphs discoveries and Implementation. In: PAKDD (2011)
38. Valiant, L.G.: A bridging model for parallel computation. CACM **33**(8) (1990)
39. Hewitt, C., Bishop, P., Steiger, R.: A universal modular ACTOR formalism for artificial intelligence. In: IJCAI, pp. 235–245 (1973)
40. Malewicz, G., Austern, M.H., Bik, A.J.C., Dehnert, J.C., Horn, I., Leiser, N., Czajkowski, G.: Pregel: a system for large-scale graph processing. In: SIGMOD (2010)
41. Salihoglu, S., Widom, J.: GPS: a graph processing system. In: SSDBM (2013)
42. Khayyat, Z., Awara, K., Jamjoom, H., Kalnis, P.: Mizan: Optimizing Graph Mining in Large Parallel Systems
43. Khayyat, Z., et al.: Mizan: a system for dynamic load balancing in large-scale graph processing. In: EuroSys (2013)
44. Bu, Y., Borkar, V.R., Jia, J., Carey, M.J., Condie, T.: Pregelix: Big(ger) graph analytics on a dataflow engine. PVLDB **8**(2) (2014)

45. Borkar, V.R., Carey, M.J., Grover, R., Onose, N., Vernica, R.: Hyracks: A flexible and extensible foundation for data-intensive computing. In: ICDE (2011)
46. Tian, Y., Balmin, A., Corsten, S.A., Tatikonda, S., McPherson, J.: From "Think Like a Vertex" to "Think Like a Graph". PVLDB **7**(3) (2013)
47. Brin, S., Page, L.: Reprint of: the anatomy of a large-scale hypertextual web search engine. Comput. Netw. **56**(18), 3825–3833 (2012)
48. Salihoglu, S., Widom, J.: Optimizing graph algorithms on pregel-like systems. Proc. VLDB Endow. **7**(7), 577–588 (2014)
49. Tasci, S., Demirbas, M.: Giraphx: parallel yet serializable large-scale graph processing. In: Proceedings of the 19th International Conference on Parallel Processing. Euro-Par'13, pp. 458–469. Springer, Berlin (2013)
50. Wikipedia. Exponential backoff—wikipedia, the free encyclopedia (2015). Accessed 13 Nov 2015
51. Wikipedia. Mean time between failures—wikipedia, the free encyclopedia (2015). Accessed 13 Nov 2015
52. Zaharia, M., Chowdhury, M., Franklin, M.J., Shenker, S., Stoica, I.: Spark: cluster computing with working sets. In: HotCloud (2010)
53. Shvachko, K., Kuang, H., Radia, S., Chansler, R.: The Hadoop distributed file system. In: MSST (2010)
54. Hindman, B., Konwinski, A., Zaharia, M., Ghodsi, A., Joseph, A.D., Katz, R.H., Shenker, S., Stoica, I.: Mesos: a platform for fine-grained resource sharing in the data center. In: NSDI (2011)
55. Vavilapalli, V.K., et al.: Apache hadoop YARN: yet another resource negotiator. In: SOCC (2013)
56. Armbrust, M., Xin, R.S., Lian, C., Huai, Y., Liu, D., Bradley, J.K., Meng, X., Kaftan, T., Franklin, M.J., Ghodsi, A., Zaharia, M.: Spark SQL: relational data processing in spark. In: SIGMOD (2015)
57. Sparks, E.R., Talwalkar, A., Smith, V., Kottalam, J., Pan, X., Gonzalez, J.E., Franklin, M.J., Jordan, M.I., Kraska, T.: MLI: an API for distributed machine learning. In: ICDM (2013)
58. Gonzalez, J.E., Xin, R.S., Dave, A., Crankshaw, D., Franklin, M.J., Stoica, I.: GraphX: graph processing in a distributed dataflow framework. In: OSDI (2014)
59. Low, Y., et al.: Distributed GraphLab: a Framework for machine learning in the cloud. PVLDB **5**(8) (2012)
60. Gonzalez, J.E., Low, Y., Gu, H., Bickson, D., Guestrin, C.: PowerGraph: distributed graph-parallel computation on natural graphs. In: OSDI (2012)
61. Kyrola, A., Blelloch, G.E., Guestrin, C.: GraphChi: large-scale graph computation on just a PC. In: OSDI (2012)

Printed in the United States
By Bookmasters